EARTHLY MATERIALS

ALSO BY CUTTER WOOD

Love and Death in the Sunshine State

EARTHLY MATERIALS

JOURNEYS THROUGH OUR BODIES' EMISSIONS, EXCRETIONS, AND DISINTEGRATIONS

CUTTER WOOD

MARINER BOOKS
New York   Boston

Without limiting the exclusive rights of any author, contributor or the publisher of this publication, any unauthorized use of this publication to train generative artificial intelligence (AI) technologies is expressly prohibited. HarperCollins also exercise their rights under Article 4(3) of the Digital Single Market Directive 2019/790 and expressly reserve this publication from the text and data mining exception.

EARTHLY MATERIALS. Copyright © 2025 by Cutter Wood. All rights reserved. No part of this book may be used or reproduced in any manner whatsoever without written permission except in the case of brief quotations embodied in critical articles and reviews. For information, address HarperCollins Publishers, 195 Broadway, New York, NY 10007. In Europe, HarperCollins Publishers, Macken House, 39/40 Mayor Street Upper, Dublin 1, D01 C9W8, Ireland.

HarperCollins books may be purchased for educational, business, or sales promotional use. For information, please email the Special Markets Department at SPsales@harpercollins.com.

The Mariner flag design is a registered trademark of HarperCollins Publishers LLC.

hc.com

A hardcover edition of this book was published in 2025 by Mariner Books.

FIRST MARINER BOOKS PAPERBACK EDITION PUBLISHED 2026.

Designed by Jennifer Chung

Library of Congress Cataloging-in-Publication Data has been applied for.

ISBN 978-0-06-304862-1

Printed in the United States of America

26 27 28 29 30 LBC 5 4 3 2 1

For E

*Que tu cuerpo sea siempre
un amado espacio de revelaciones.*

*May your body always be
a beloved space of revelations.*

Alejandra Pizarnik

Contents

Preface xi

I. Mucus *1*
II. Urine *25*
III. Blood *37*
IV. Semen *73*
V. Menses *107*
VI. Milk *133*
VII. Flatulence *175*
VIII. Breath *201*
IX. Feces *221*
X. Vomit *239*
XI. Hair *299*
XII. Tears *321*

Epilogue *351*

Acknowledgments *361*

Preface

An hour's drive from present-day Baghdad, a series of mounds dot a desolate reddish plain. Four thousand years ago, before the course of the Euphrates migrated west, this was the ancient Sumerian city of Nippur. We know from a series of archaeological digs in the late nineteenth century that the settlement housed a large school for scribes. There, along the river, young men and women learned what was then the relatively new art of writing. From the shallows, they gathered reeds, cutting them obliquely to create their styluses. They collected clay from the river's banks and molded it into tablets or flattened circles or prisms, and by pressing their sharpened reeds into this still-wet clay, they created the uniquely angular and cone-shaped script that gives their writing its name: cuneiform.

Sumer was an association of city-states, the largest probably containing no more than eighty thousand people, about the population of a handful of blocks in Manhattan. They built temples of mudbrick, went to war in chariots, played

PREFACE

bull's head lyres, and worshipped a panoply of gods and goddesses with names like An and Ninhursag. In a scribal school like the one at Nippur, students learned to write, in part, by copying out well-known sayings, and much of what we know about the civilization comes from the study of these proverbs.

Most of the proverbs have the distinctive smack of a small farming city on an ancient plain. There is a kind of homesteader's wit and wisdom: "There is no baked cake in the middle of the dough." There are also aphorisms that seem to capture the specific violence of life in that era: "You should not cut the throat of that which has already had its throat cut." Most are deeply dated, and many are simply confusing: "What is eaten for today was put there by the dog. What is eaten by the dog was put there for today." The twelfth proverb, though, stands apart from the rest. It invokes no vanished deities or lost cultural customs. Nearly two millennia before the birth of Christ, a student in Nippur pulled a lump of clay out of a basin of cool water, flattened it into a circle, and wrote these words:

> *Something which has never occurred since*
> *time immemorial: a young woman did*
> *not fart in her husband's embrace.*

It's the world's first recorded fart joke, but what's funny about the twelfth proverb isn't the joke. Neither is it that humans should have evolved the faculty of language, then of writing, only to set down what is little more than a crude knock-knock. What's funny is that somehow, even thousands of years later, it feels like a joke we might still make. In fact, it is. There it is in season two, episode 17 of the sitcom *3rd Rock from the Sun* when Mary, played by Jane Curtin, nonchalantly passes gas in front of her boyfriend (John Lithgow), an alien masquerading as a suburban father. It surfaces again in season one, episode 11 of *Sex and the City* when Carrie (Sarah Jessica Parker) does the same while snuggling with Mr. Big, thus throwing their affair into one of its many tumults. And there it is again in *Mallrats*, as Brodie worries his flatulence may have been the death knell for his relationship.

We now govern through an institution of representative democracy, we send ships careening through space at 165,000 mph, we snip and sew the very genetic ribbons from which life is composed. I write these letters down for you not with a sharpened reed yanked from the Hudson but by shuffling countless subatomic particles through a series of infinitesimal electronic switches. And yet the basic functions of the human body remain so fundamentally confounding that we, as a culture, are still telling a four-thousand-year-old fart joke.

ANY BIOLOGY TEXTBOOK USUALLY FEATURES A DIAGRAM OF the human body flayed to reveal its complex inner workings—the red and blue pipes of the arteries and veins, the intricate coils and loops of the reproductive organs, the endless branching roots of the nervous system. We learn that the body is a fundamentally cohesive organism, a collection of organs and tissues, fed by the heart and the lungs, overseen by the brain, wrapped up in a covering of skin, the whole shebang working together to a common purpose. But while this idea of the body isn't false, it's deceptive. The human organism depicted on the page fails to give us a complete picture in one very crucial way: it looks tidy. In reality, it is anything but. Whether it is blowing its nose, combing its hair, emptying its bladder, or simply exhaling, the human organism is essentially permeable. It leaks.

To live, our bodies must continuously shed material. Eat last week's potato salad, and the stomach expels the offending material. Inhale a virus or a piece of pollen, and the nose runs. Digest a fiber, make a muscle, send a signal from one end of a neuron to the other, and you create a chemical byproduct that must be reprocessed or removed from the body in its own specialized manner. There's nothing we can do about it. Stop urinating, stop defecating, stop breathing, and death is near.

We can learn a great deal about ourselves from these materials. The color of our urine, the volume of our flatus, the frequency of our tears, the rhythm of our breath: taken together, these materials tell a story of the human that produced them, and it's curious then how hesitant we are to acknowledge, let alone discuss, them. Instead, we often hide our body's productions.

We cough into our elbows and hurriedly plunge the recalcitrant stool and stash the sodden boxers and turn our heads to cry. We act as if the mention of a bowel movement at the dinner table will rock the Earth on its axis, and I sometimes wonder if this prudishness about our bodily materials stems in part from an intuition that the stories they tell might diverge from those we would prefer to believe. It's far easier philosophically, and far safer emotionally, if we consider these materials mere byproducts of the real work of being human.

But we're wrong to do so. Whether we acknowledge it or not, our shedding bodies put us in constant dialogue with our environment and our fellow human beings, and in so doing, these materials play a central role in who we are and how we organize ourselves in society. The architecture of our homes is determined by the need to safely dispose of excrement, and our cities, perhaps the single most notable feature of the human experiment, are only made possible by a vast infrastructure of sewage disposal. Our mucus forms a genetic bazaar where both dangerous and useful microbes are continuously traded back and forth with other organisms. We sell our blood and tax our tampons and boil our urine and say ten Hail Marys as penance for our masturbation. Our very breath is political.

The exchange, elimination, and frequent disguise of our effluence, in other words, has been elemental to the development of human civilization, and our lives today are still governed by a host of laws and mores, superstitions and misconceptions, about the materials our bodies leave behind. When we breathe, when we vomit, when we urinate, we negotiate an uncomfortable and often untenable truce between our own body and society.

Each chapter in this book is about one of these negotiations. In these pages, you will find an alchemist hoarding his urine, theologians debating the resurrection of the intestines, a museum committee flummoxed by nine kilograms of hair, a church consecrated by vomit, and several hundred thousand young men trying not to masturbate. You will also find an infant formula crime syndicate, a mucus centrifuge, and several weeping philosophers. I have not provided a grand and cohesive argument about humanity here for the simple reason that the human body is neither grand nor cohesive. Rather, in twelve stories, I describe a kind of terrestrial organism. It is a peculiarly social creature. Large-brained, bipedal, it lives in groups on a smallish watery planet with an oxygen-rich atmosphere. It gives birth to live young and feeds them with liquid exuded from its own body (or often with the powdered milk of another species). It urinates and defecates in closets and blows its nose in moisturizing tissues. It is transported by the scent of hair or the delirium of nausea. It absorbs its disintegrating endometrium with lozenges of cotton cellulose, ejaculates to two-dimensional recordings, plays games with its flatulence. It cries and wonders why. It's a messy organism, in other words, and so it is a messy book.

That mess itself is my goal. If you leave these pages thinking you understand *Homo sapiens*, this book has failed. But if one day you inspect the toilet bowl before flushing or look into the tissue after blowing your nose; if you take a moment to smell the hair of a loved one or to listen to the pitch of your own flatulence; if you try to find the seat of your nausea or the moment when a tear becomes inevitable; if you pause and consider the materials you are forever leaving behind, that will be a beginning.

I

Mucus

Subject: A perhaps unknown property of mucus!

I wanted to tell you about an attribute of mucus I discovered by accident.

To the point, somehow (I sneezed or something), I got mucus or snot onto an exposed minor scratch which was still bleeding slowly. I got distracted doing something, and when I could give it my attention it had scabbed over. It looked odd and as I fooled with it I found that it seemed several times stiffer and stronger than a blood-alone scab. I left it alone and the scratch seemed to be healing quickly without inflammation.

I think it's interesting and might be worthy of research to see where it leads.

I am writing this message to see if your work provides any insight into the creation of thickened mucus (particularly in the throat) and whether there might be some concepts for reducing this thickening.

i wanted to pass on a reference to a mask and society of the NW kwakiutl tribe that studied mucous circa 1840. Always good to know context . . . the smithsonian has their number reference and you might get more info from them. if you do find out more, copy me

Is it bad to eat your boogers?

Inquiries received by MIT's Biogel Lab in Cambridge, Massachusetts

Terms

n. snot, boogers, boogs, gold, loogie (when hawked), phlegm, nose fruit

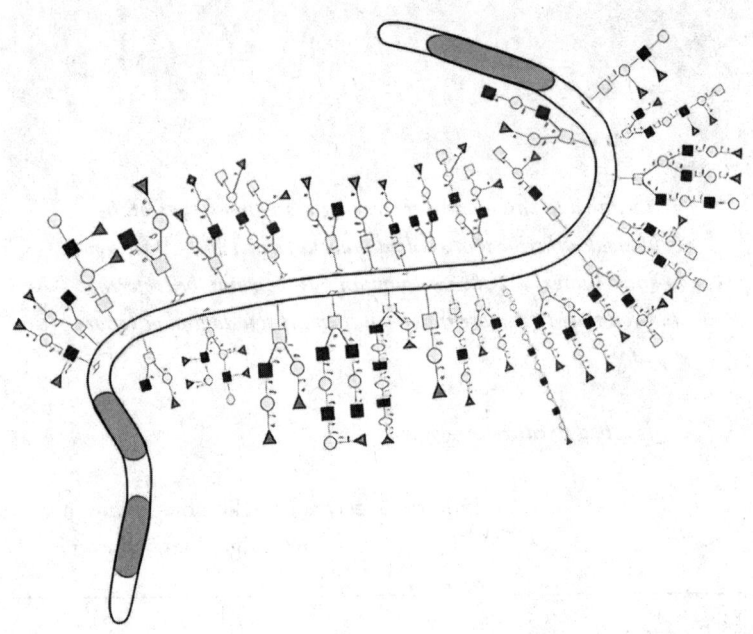

Diagram of a mucin, "A Complex Connection Between the Diversity of Human Gastric Mucin *O*-Glycans, *Helicobacter pylori* Binding, *Helicobacter* Infection and Fucosylation," Gurdeep Chahal et al., 2022

Biological Prologue:
The Multiscale Organic Mesh

Things are always going up your nose. It could be a single bacterium wafting through the pressurized atmosphere of an airplane cabin, or after visiting a desperately sick aunt, you might rub your nose absentmindedly and leave behind the protein-spiked sphere of a rhinovirus. Or some afternoon at the botanical garden, you might bend down to sniff a rose and inhale along with that ethereal carmine scent a golden clump of pollen. That golden clump or spiked sphere or bacterium, during its brief adventure in the nasal passage, careens among hairs, is jostled by eddying currents of air, and finally lands, not on the body's epithelial cells—and this is of the utmost importance—but on the layer of mucus covering those cells, and there the thing sits, trapped, like one of those bugs in Jurassic amber.

Depending upon the physical and chemical nature of that foreign body and upon the sensitivities of the particular nose into which it's been introduced, a number of events might occur upon arrival. The object might be permitted to penetrate the barrier and reach the epithelial cells below, as is the case with certain chemicals, or it might be encased in thicker varieties of mucus while elements of the immune system mobilize to fight it. It may attach itself to sugars in the substrate and be sheltered indefinitely in the mucosal layer, or, as is often the case, the immune system could raise its already robust rate of mucus production, causing the nose, as we say, to run.

Mucus is produced continuously by the submucosal glands

and by the satisfyingly named goblet cells. It is created in the nose, the lungs, the eyes, the digestive tract, the reproductive organs, among other locales. On an average day an average human body makes a great deal of the stuff, something on the order of a gallon—enough that the combined efforts of all humanity could fill ten *Titanic*s to brimming every hour. It's composed of water, salts, fats, sugars, suspended in a gel formed of the long, thin, and eponymous protein mucin (imagine a strand of fettucine covered in dust); each of those strands of fettucine forms a variety of bonds at different sites on other strands, creating what is sometimes described by scientists as a complex multiscale organic mesh.

Because that mesh traps things, and because its texture determines what it traps, our mucus can tell us a great deal about ourselves. White or clear indicates healthy production of mucus, while mottled yellow or green signals the presence of white blood cells and the likelihood of an infection. Puce or scarlet (or even sometimes a smooth and glasslike obsidian) is a sign of blood in the mucus, whether through superficial laceration or from some other deeper and more foreboding source, while a grayish tinge provides good evidence of exposure to environmental pollution. Its consistency, meanwhile, testifies to the environment in which it formed, with drier conditions producing the desiccated mucosal product known in English as a booger (perhaps from bogey, Scottish *bogle*, "a phantom or goblin," or from bugger, Old French *bougre*, "a heretic;" the etymology is as maddeningly inviting as it is indefinite). This isn't merely funny–ha ha material. These are actual diagnostic criteria. Slight variations in the texture of mucus, what scientists call its spinnability, can be used to diagnose serious medical conditions; the spinnability of a pregnant woman's cervical mucus, for instance, is one of the better ways to predict her likelihood of a miscarriage.

On a practical level, everyone knows what the stuff that comes out of their noses is. We do not call it a complex multi-scale organic mesh; we call it snot, and we know exactly what to do with it. We blow our noses, get rid of the results, and go about our lives. In this moment of minor annoyance, our primary experience of mucus, all our thoughts are summed up with the swipe of a tissue. But there's more to discover, as every child recognizes when they put their finger up their nose. It's all a matter of how far you're willing to go.

Borderlands

When I make the trek to MIT's Biogel Lab, there's a lot going on in the world. Protests continue in Hong Kong. In northern Syria, Abu Bakr al-Baghdadi, the de facto leader of ISIS, detonates an explosive vest as United States Special Operations forces surround him, and the president, trying to convey to reporters the feeling of watching the death via live video feed, says it was like "watching a movie." In the past twelve months, more than seventy-five thousand children have been taken into custody at the U.S.-Mexico border, and in California, wildfires have turned the outskirts of Los Angeles into a charred and smoking wasteland. The smoke is so thick, the red carpet premiere of *Terminator: Dark Fate* has been canceled.

There's the creeping sense, in other words, that the web of human connections has begun to rapidly unravel, but this fades away when I arrive in Cambridge, Massachusetts. The day is overcast but brisk. The Charles River looks like quicksilver. On MIT's campus, construction crews are busy knocking down one building—the former Wright Brothers Wind Tunnel—and putting up a half dozen others. A tour guide is explaining to a group of prospective students and their parents the sort of research that will one day be conducted in these buildings. Progress here feels as real as the steel girders waiting to be lifted into position.

The Massachusetts Institute of Technology, along with its reputation for academic excellence, is also noted for its dedication to meritocracy, its tradition of elaborate practical jokes, the role it's played in the computer revolution, and its outstanding model railroad club. If these data points don't provide enough detail on the sort of pimpled, kyphotic anticharisma—itself somehow charming—that pervades MIT's campus, then I refer the reader to a T-shirt once popular among the undergraduates of nearby Wellesley College: MIT, *the odds are good, but the goods are odd*. In the hubbub of students hurrying to grab a bite between classes, one student wears a sweater vest, another has two buns of auburn hair on either side of her head in the mode of a certain distressed fictional space princess. Bits of conversation filter out of the fall air like lines struck from an episode of *The Big Bang Theory* for being too nerdy:

"You know ice-covered settings preserve more information."

"So are the NACG research goals the same as for the NSF proposal?"

In the middle of it all, untouched and alone on a perfectly smooth sidewalk, a young man on a bicycle abruptly crashes. A man raking leaves shakes his head and goes on raking.

Behind a neoclassical hall with towers dedicated to men like Linnaeus, Newton, and Faraday, away from the vast lawn of Killian Court and the slow churning of the Charles, there's a misshapen quadrangle composed seemingly of the backs of several buildings. Asphalt and concrete predominate. A few industrial-size tanks of argon and liquid nitrogen hiss. This is Building 56. This is where a cadre of engineers, biologists, and graduate students, led by Dr. Katharina Ribbeck, has spent the better part of the last decade trying to understand precisely what mucus is and does.

I've spoken and emailed with Dr. Ribbeck a few times in the run-up to this visit, and in the process, unwittingly, I've gotten some image into my head of what a mucus scientist might look or act like—earrings of eccentric proportions, erratic yet voluble laughter—but when I find her office on the second floor, I discover Dr. Ribbeck is wholly out of line with nearly all of my preconceptions. All in black, with steady eye contact, a disarming smile, and a firm-but-not-too-firm handshake, she has more than a little of the politician's proprioceptive abilities. She inquires about my travel, did I have trouble finding the lab (of course not: she has previously sent detailed directions, a map, and her cell phone number), would I like water or perhaps coffee. She gestures toward her desk, where a coffee cup of double-walled borosilicate glass bears the distinctive rime of a cappuccino. It is the intellectual savoir faire more commonly associated with villains in James Bond. As we take our seats at a table, a black poodle, silken and aloof, enters the room and settles down in the middle of the floor.

The art on her walls appears to be something in the vein of Jackson Pollock, and as we exchange formalities my eye is drawn to the pattern of loops and spirals. The more I look at it, the more it seems to resemble one of those Hubble stills of a distant star cluster where wisps of material cohere into swirling patterns of nodes and cavities, but the density here is not quite stellar. I lean in closer, but I can't put my finger on exactly what I'm seeing.

"Mucus," says Ribbeck from across the table. "At five hundred times magnification, I think."

Dr. Ribbeck speaks English as a second language with more precision than I do as a native speaker, and she has an acute manner of listening, where you can tell that not only has she already parsed out your whole sentence even though you're only halfway through speaking it, she's already preparing her own response. One of my first questions has to do with the recent popularity of mucus as a subject of research and with the long period of neglect that preceded it. She doesn't miss a beat.

"Speaking for mucus," she says.

This is what she does day in and day out, after all, expressing to others what mucus can't explain for itself. But it's also one of the phrases that *no one ever hears*, and it is thus utterly delightful.

"I have a soft spot for these gels," she says.

Ten years ago, Ribbeck tells me, when she first proposed mucus as a worthy object of study, half the faculty of Harvard's biology department cringed, but time has been kind to mucus, and these days the stuff is very much in the zeitgeist, in no small part thanks to the work done at this lab. In rapid succession, she reels off a few talking points. As the first line of defense in the body's immunological response, mucus is tied up in the etiology of many allergic reactions, and is implicated in a vast array of ailments from the passing hay fever of a June day to more severe and complex lifelong diseases such as Sjögren's syndrome, where a failure of mucus production in the mouth and eyes causes the eyelid to stick to the eyeball, as she describes it, like Velcro. With its antimicrobial properties, mucus is at the center of discussions about a postantibiotic world, and in its role as an efficient and selective physical filter, it holds promise for an array of both pharmaceutical and industrial applications. Perhaps foremost, as the soil in which the body's microbial community grows, mucus finds itself involved in the trendiest of fields, the reimagining of the human organism less as an integral unit than as something with much fuzzier edges, and it has begun to take its place in the investigation of the many ailments—from autism to gluten intolerance to ADHD—that have been attributed, rightly or wrongly, to deformations of the human microbiome.

Mucus, in other words, is hot right now. As I've discovered, once the stuff is on your radar, it's hard not to find muc-news cropping up everywhere. The mucus of the Chinese giant salamander heals wounds, while the mucus of Israeli jellyfish filters microplastics from polluted seawater. Fish mucus, or sometimes snail mucus, holds the key to overcoming antibiotic-resistant bacteria. Whales near Australia are being followed by helicopter drones that sweep down to collect the mucus particulates ejected when they expel air from their blowholes, while the whales off Alaska are being tailed by mucus-collecting submersible drones. Swiss scientists are smearing francs with mucus containing the human influenza virus.

"Mucus is experiencing a renaissance," Ribbeck says. "It's a very sophisticated material. You couldn't design it if you wanted to."

As much promise as this research holds, the work done at the lab is still, in Ribbeck's words, "material-limited," which is to say that despite the natural abundance of the stuff within the human body, it's hard to get quality mucus, and many of the experiments are designed with this deficit in mind, whether it's testing viscosity via disturbances of a microscopic needle or determining mucoadhesivity by pulling apart two parallel plates between which has been sandwiched a single drop of mucus.

"Getting mucus is not super convenient," she explains as she walks me down to the lab proper, "but it's not impossible." They can always order dehydrated mucus from the biomedical supplier Sigma-Aldrich in Milwaukee. (As can you: M3895, mucin from bovine submaxillary glands Type I-S; M1778, mucin from porcine stomach Type III. But it isn't cheap. A gram of M3895 will set you back $499.00.) They prefer to produce their own lab-quality mucus. "The biggest obstacle to getting mucus," she tells me, "is conceptual." One of their two sources is a man who drives down from the slaughterhouses in New Hampshire with a cooler of animal parts in his trunk. He's something of a fixture in the Greater Boston Area, making the rounds of various hospitals and universities, delivering eyeballs here, livers there. For the Biogel Lab, he saves the stomachs and intestines. Dr. Ribbeck opens the door to the lab and gestures for me to follow: "They have prepared for you a stomach, I believe."

THIS IS WHERE BRAD TURNER, THE LAB'S CHIEF TECHnology officer, enters the picture. Though the processes for purifying mucus have been around since the 1950s, Turner, a biochemist by training, has been perfecting the technique here at the lab. An older man, he wears a plaid shirt, the breast pocket nearly bursting with pens, and a white lab coat. Fine, tightly curled hair, mottled gray like a storm cloud, covers his head and cheeks, framing a wide face, a pair of soft, brown eyes, and the permanently furrowed brow, brooding yet pleasant, of a man who has spent much of his life considering problems on the molecular scale. Despite all the advanced equipment in the lab, mucus purification begins with nothing more than a cookie sheet, a bus tub, two paint scrapers, and forty pig stomachs. The stomachs arrive still full of the pigs' final meals, and the first step, he tells me, involves emptying the organs of those contents.

"The smell can be pretty unpleasant," Dr. Turner says, using a gloved hand to scoop out a grainy oatmealish substance that sticks to his glove so tenaciously he must give a little shake of the wrist to get the material to fall into a trash can. As he sets to work in earnest, he takes on the manner of a professor, filling the silence with the tangential minutiae that are second nature to a lecturer. "Pigs are partially copraphagic," he says, scraping down the stomach. "They rely on nutrients [*scoop*] that are only made [*shake*] when bacteria digest their feces, so [*scoop*] they need to eat that material to get those nutrients [*shake*]." He turns to me then, gesturing vaguely with the pig stomach in one hand. "Guinea pigs and rabbits are the same way. They'll die if they're unable to . . ." He stares down at his paint scraper—a ruffle of gelatinous brown muck is stuck to the tip—and trails off.

From forty pig stomachs, they get about 100 to 200 mL of what Turner calls crude mucus. Shedding the lab coat and peeling off the gloves, he turns his back and sets off into the lab, gesturing to the tables and machinery we pass. They dissolve the gel overnight in two liters of salt water, he says, then they run it through what looks like a monstrous Soviet-era washing machine-and-dryer combo. They're centrifuges, one that spins at ten thousand times the force of gravity, the other spinning at fifty thousand times. Then they filter it through a G-25 fractionating column using size exclusion chromatography.

"Mucins have a high molecular weight," he says. "Hemoglobin weighs something like sixty thousand daltons, but mucus is several million, so . . ." He trails off again and heads farther into the lab. "Then we dehydrate it." He opens a drawer and pulls out a small ziplock bag containing what appears to be a tuft of cotton candy. The whole process takes about a month, and the forty stomachs will end up yielding about a gram of the stuff. He holds the bag in an open palm. "We calculated what it was worth at one point," he says, allowing himself a nervous smile. "Something like ten thousand to twenty thousand dollars per gram."

THE BIOGEL LAB DOES HAVE ONE OTHER SOURCE OF

mucus: native Cantabrigians. Mostly made up of the lab's doctoral students and their friends, the mucus donors come for about an hour at a time to sit in a room off the office while a breast pump drains their salivary mucus into vials. I have been asked, somewhat sheepishly, first by Dr. Ribbeck, then by Dr. Turner, if I would be interested, as long as I'm in the lab, in donating my mucus to the cause of mucus science, and after I've been walked through the purification process, I am placed in the care of three eager PhD candidates, who usher me into a small room.

"We call it the small room," one says. "We're not very inventive."

The four of us discuss O-type bonds and sialic acid while they get the apparatus ready. One gives me cotton wadding and counsels me on how best to stuff it in my cheeks. Another disappears and returns with a red Solo cup of chipped ice. The third cuts a length of plastic tubing from a large spool and runs it between a nozzle on the breast pump and a vial sitting in the chipped ice. As we talk, I'm overtaken by a familiar feeling. I've been trying to put my finger on it this whole time, but only now does it start to come clear. What has impressed itself upon my senses at the Biogel Lab isn't the high-tech machinery, the cooler of pig parts, or the aggregation of advanced degrees. It's the level of enthusiasm about mucus. There is a vibratory electrical charge in the air that reminds me of my childhood elementary school in the throes of an arts and crafts project. Replace construction paper and Elmer's glue with a breast pump and tube, and you have the general feeling of what it's like to sit in the small room at the Biogel Lab and talk about mucus.

In fact, you don't talk *about* mucus at all, you talk mucus, as in "We love to talk mucus" or "I could talk mucus all day." Mucus is the métier. These people have chosen to devote their lives to a substance that many human beings, if not most, find actively repellent. They come in early so they can sit in this room watching cooking shows on YouTube ("I like Jamie Oliver," one says) and have their bodily fluids drained off. And while that sort of time investment and countercultural attitude almost inherently skews your perspective about the importance of a subject, having only been in the lab a couple of hours, I am already nodding along enthusiastically and asking them to repeat exactly how many kinds of mucus there are in the human body (at least twenty-one) and what the difference is again between MUC5AC and MUC5B and how exactly the glycans are sequenced.

By this time, the machinery has been rigged. There's just one final piece of plastic tubing that needs to be attached. One of the grad students holds it up, and only when she places it in my hand do I realize this is the part that connects to me. It goes under the tongue, she explains, already edging toward the door. Another of the students, in the moments before I am left alone with the pump and the tube and my thoughts, confides in me that she always looks at pictures of pickles.

This room too features a wall adorned with an extravagantly magnified image of mucus, and as I settle into my beanbag chair, my cheeks puffed with cotton wadding, and stare up at the poster, it strikes me that mucus, once you start to think about it, really is pretty wild. There's a fractal complexity to the stuff that invites a stonerish zoning in of the gaze. Part of this must derive from the fact that, though everyone now recognizes the nearly mystical operation of an organ like the brain, the general public hasn't got any clue just how mysterious snot can be. The other part of the excitement is that, as with brain research, the more we learn about mucus, the more we realize just how little we understand.

Take, for instance, the single function that most people recognize as the purpose of mucus, the role it plays as a filter. Dig out your most recent high school biology textbook, and you will almost certainly find a description of mucus acting as a physical mesh, basically a series of nets designed to seine particles of a certain size. "Wrong," says Dr. Ribbeck. If you were to proceed with your research, delving into more recent literature, you might find that a molecule's net charge or ability to dissolve in fat will determine its ability to pass through mucus, but this too turns out to be, if anything, only a small part of the picture. The simple fact is that, except in a very piecemeal way, we still don't know how mucus filters out certain things while allowing others in. The best we can say, and I'm paraphrasing here from the impressively dense *Annual Review of Cell and Developmental Biology*, is that the filtration has to do with the spatial distribution of charged and hydrophilic groups along the overlapping mucin strands, which distribution may be altered by shifting the pH balance, and since next to nobody really gets that, we can mostly just say that mucus is a very, very good filter.

But even that sort of hole in our knowledge, while substantial, is dwarfed by comparison to what we don't know about mucus's other main function. As is becoming increasingly clear, perhaps the more fundamental purpose of mucus has to do not with the microbes it keeps out but with those it keeps in.

STAPHYLOCOCCUS AUREUS IS A ROUND GOLDEN BAC-terium that, when viewed under a microscope, often appears in clusters similar to bunches of grapes. In humans, *S. aureus* is the most common source of bloodstream and joint infections. It can cause boils and impetigo on the skin, and if it navigates its way to the heart, it can dissolve the organ's valves. In the gut, it sometimes secretes staph toxin as a way of destroying potential competitor bacteria, causing food poisoning in the process, and on implanted devices such as pacemakers, catheters, stents, and penile and breast implants, it creates a biofilm that makes it impervious to antibiotic treatment and requires the removal of the device. In chickens, it causes bumblefoot, mastitis in cows. The list goes on. It is, in other words, a highly virulent pathogenic bacterium. It reproduces quickly, and when allowed to spread unchecked, it can kill a human in a matter of days, which is why it was one of the first to be treated with penicillin, and likewise why it was one of the first to show signs of antibiotic resistance. *S. aureus* is now one of the most common hospital-acquired infections after surgery, and some 50,000 deaths are linked to it annually in the United States, roughly one in 7,500 people, or the equivalent of one Yankee Stadium at capacity kicking the bucket each year. You may know someone who died of a staph infection. I do. And yet, it's estimated that 30 percent of the population carries this bacterium in their noses, on their skin, and in their gastrointestinal and reproductive tracts without showing any sign of bodily distress. They live with it, and it lives with them.

The other function of mucus is its role as a substrate for one of the densest ecological communities ever documented, the body's microbiome, a living colony of diverse bacteria. "Mucus contains the body's largest reservoir of complex sugars," according to Dr. Ribbeck. "It is the soil for the microbiome." It is this ability of mucus to house and feed bacteria that sends certain graduate student mucus researchers into gyratory hand gestures, soul-defining stares, and mutterings about "glycan motifs" and "ablated pseudomonas behaviors." Because for microbiologists, biophysicists, computational biologists, statistical physicists, and game theorists, the term "microbiome" carries with it the sort of anticipatory and ominous tone of a distant clarion. Anticipatory because of the rapidity of recent discoveries, ominous because the scale of what remains to be done is daunting in about ten different ways.

Far from the solitary cells so often witnessed swimming across microscope slides, the bacteria in the microbiome, bacteria like *S. aureus*, turn out to be complex creatures, capable of forming highly organized and efficient (and usually sticky) communities known as biofilms (scrape your teeth up by your gum line and you'll find one). And the relationships between different bacterial species' biofilms can be contentious at best. Grown in a petri dish of agar gel, the various species of bacteria in the human microbiome will rapidly alter their dispositions, taking on a distinctly competitive aspect. Some will grow tails, others will secrete toxins that are deadly to their neighbors. This arms race can rapidly escalate into an all-out bacterial war, and in short order, one or two of the species can monopolize the available sources of energy, forcing the others out of existence and making life generally unpleasant for any cells nearby.

It is precisely those virulent traits that make *S. aureus* so deadly, and yet, through a variety of means that are only just becoming clear, mucus somehow tames these behaviors, not just of *S. aureus* but of a community of microbes potentially numbering in the thousands of species. Via the sugars studding its protein backbone, mucus feeds specific microorganisms in such a way that their more virulent functions are suppressed. As for *Candida albicans*, the fungus responsible for thrush, mucus makes it take on a novel form, preventing the expression of genes related to the formation of filaments or the attachment to surfaces. It also prevents the communication of chemical signals that cause bacteria to form biofilms, and when a bacterium does become virulent, attempting, in the case of *Helicobacter pylori*, to implant itself upon the surface of the gut, mucus lures it with a unique glycosylation pattern, binds to it, then detaches and carries it away.

The magnitude of this feat of peacekeeping should be clear to anyone who has ever lived in a dormitory, been subjected to episodes of *The Real World*, or read a history of medieval Europe. It is not easy for multiple beings to coexist in proximity, and the fundamental role mucus plays in maintaining bodily harmony is evidenced by what happens when your mucus stops working. According to the NIH, an estimated 80 percent of all internal infections are related to mucosal dysfunctions.

Why does the body go to all this trouble of keeping bacteria alive in the first place? The advantages are clear, as many studies have confirmed. The greater the diversity of the body's microbiome, the healthier the human. Many of these bacteria offer some advantage to the body. *Lactobacillus casie* and *Lactobacillus plantarum* team up to digest the fibers in beans, turning them into a form of energy the body can absorb (and producing as a byproduct methane). *Clostridium sporogenes* synthesizes indolepropionic acid, a neuroprotective antioxidant. Many produce the short chain fatty acid butyrate that then becomes fuel for cells in the gut. Less clear is exactly how mucus gets these bacteria to peacefully coexist. The number of species in the average human microbiome hovers around ten thousand. Using traditional mathematical modeling, a network of just 170 species of bacteria would have more possible interactions than there are atoms in the visible universe.

THE BODY LEAKS IN ABOUT A DOZEN DIFFERENT WAYS, give or take a few, depending on how finicky you're being about definitions, but mucus might be the most challenging to wrap your mind around. Urine, feces, breath, these are almost purely wastes; blood, on the other hand, is an undeniable part of the body. But mucus is somewhere in between. Mucus *is* the transition between the body and everything else, and the question then that's all but unavoidable as you sit in the Biogel Lab looking at pictures of pickles on your phone and thinking about bacteria while your mucus drains into a vial: Does this stuff qualify as *me*? Is mucus basically just this goo that coats us? Or is mucus a part of us, but the myriad junk living within it is not? And if we go that route, how do we explain the huge amount of bodily work that goes on in the mucosal layer, the metabolization of countless chemicals necessary for our continued existence? Or do we come at this from the other direction and lump both mucus and the microbiome in with our definition of self? And if the microbiome is me, what does that mean?

As it turns out, among evolutionary biologists there's actually a rather intense debate right now about this very question, a debate that has rapidly differentiated the involved biologists into two entrenched and opposing camps. One group espouses what we might call the standard model of natural selection. This holds that an individual in a species is composed of their own genetic material. Random mutations in this genome confer competitive advantages on the individual, allowing those genes to be passed on, so that other less useful genes fade away into the fossil record. Imagine, for instance, that your great-great-great-great-grandfather had a mutation that allowed him to digest potatoes. This was a significant advantage because it opened up an entirely new source of energy to him that wasn't available to others. He flourished, having absolute scads of offspring and passing along to some of them his potato-digesting gene. Those offspring in turn flourished, scads, children, flourished, scads, potatoes, and so on and so forth. This is your basic survival-of-the-fittest example of natural selection. It's how humans got thumbs and roses got thorns.

As the importance of the microbiome has come into focus, however, a group of evolutionary biologists has begun to argue that we should be looking at more than just the genes of an individual. This camp contends that the well-being of a human and their bacterial entourage is so deeply interwoven that we need to create a larger and more comprehensive evolutionary unit that includes not just your own genome but also the genetic material of your entire microbiome. In the case of your grandfather, suppose instead that the capacity to digest potatoes didn't actually belong to him but to some bacteria that happened to find their way into his digestive tract. The argument of these folks is that we should consider your grandfather and his bacteria *together* as a distinct evolutionary unit. When bacteria in our guts help to digest our food, when bacteria in our sweat glands help to communicate signals about fitness for mating, these scientists argue, we should be considering the evolutionary unit not really as the single organism but as the collective of interconnected organisms. They even have a name for this collective. They call it the holobiont.

This new mucus research doesn't necessarily decide the argument in favor of the holobiont concept, but it suggests, at the very least, that our evolution has been far more complicated and collaborative than we once believed. As Seth Bordenstein, a biologist at Vanderbilt University, has put it, "If host-encoded gene products such as mucin select for or against certain microbes, then host evolution impacts holobiont composition. That would be further evidence for hologenomic selection." In other words, if it turns out that your grandfather's mucus specifically selects a bacterium from the environment and feeds it, that's a strong argument for considering them as a single unit.

A shift like this makes our ideas about evolution and about competition much more complicated. You don't just get bacteria from your parents, after all. You mainly acquire them from the environment, from that breath of air you just took, from the banana you had for breakfast, from the guy at the deli who made your hoagie. If you're continuously selecting and feeding and manipulating a phalanx of additional genetic material from your surroundings, that means you're a lot less *you* than we usually think. You would be constantly giving pieces of yourself away and receiving pieces of others. Life, both as moment-to-moment survival and as the long evolutionary arc, would be much more collaborative and entangled than we currently think of it.

It's been 160 years since Darwin debuted the theory of natural selection, and in that time, the idea that we are all fundamentally discrete and selfish actors has been absorbed into everything from xenophobic political sloganeering to free market economic modeling. To think that a soldier might return home to Idaho from Afghanistan as a genetically different organism and might lend those genes to people in her community; to think that a child in the custody of Border Patrol is already irrevocably becoming a citizen of a new country; to think that a complex multiscale organic mesh, mostly encountered in the vicinity of Kleenex, has the potential to upend an entire way of viewing the world—that's bound to give a person pause. Here is mucus, a sort of genetic bazaar where we swap actual functioning pieces of ourselves back and forth. Here is mucus, this peculiar borderland where antagonistic entities from the exterior world are tamed, turned into cooperators, and incorporated into a dynamic functioning whole. I get the rapture these scientists must feel when they hold a vial of purified mucus up to the light. If we ever do learn how those thousands of species manage to cooperate in the dark hold of the human body, maybe it will give us a clue or two about how we as humans can get along.

II

Urine

Our eye asks our urine how we do.

John Donne, *Devotions upon Emergent Occasions*

> **Terms**
>
> n. piss, pee, peepee, whiz, wee, wee-wee, leak, water, piddle, tinkle, sprinkle, number one
>
> v. take a leak, take a whiz, drain, empty, lift a leg, void the bladder, make water, break the seal, write one's name (the Keatsian phraseology) in the snow, hacer pipí

Diagnostic urine chart, *Epiphanie medicorum*,
Ulrich Pinder, 1506

Biological Prologue: Urinalysis

Over the course of any given day, a person usually evacuates somewhere between one and two liters of urine, though that quantity can be influenced by a variety of factors, from fluid intake to humidity levels to medical conditions and the drugs prescribed to treat them. The average quantity voided in any given session and the frequency of these voids are widely variable and difficult to assess, given the vast number of confounding factors. Some go more than six hours between voids, while others must urinate before two hours have passed and rise throughout the night, as well, to make their pilgrimage to the toilet. Broadly speaking, though, women tend to void less urine than men and do so more often, and everyone tends to urinate more frequently as they age and their organs become less efficient. While you can't divine every malady from urine, as some medieval uroscopists believed, urine output is indicative of the health of the kidneys, bladder, and urethra, and a number of serious medical conditions—obesity and hypertension, to name two—are associated with frequent urination and nocturia (urinating more than once at night).

Urination also has a psychological component. Overactive bladder, the frequent feeling of a need to urinate, has no exact definition or pathophysiology, but it is associated with stress and anxiety. Upon being deployed to serve in the war in Iraq, for instance, many soldiers reported experiencing, alongside a cornucopia of mental health issues, a frequent need to urinate. And George Orwell recalled his first days away from home on scholarship at a boarding school and how his monumental struggle not to wet his bed in the night, and the beatings he received from the headmaster for these bed-wettings, cemented in his young mind a distinctly dim conception of the world:

> It was possible, therefore, to commit a sin without knowing you committed it, without wanting to commit it, and without being able to avoid it. Sin was not necessarily something you did: it might be something that happened to you . . . this was the great abiding lesson of my boyhood: that I was in a world where it was *not possible* for me to be good.

Urine is slightly acidic, slightly denser than water, and, contrary to popular belief, far from sterile. The composition varies from person to person and moment to moment, but it's mainly—about 96 percent—water. Much of what's dissolved in this water is the body's excess nitrogen, excreted in the form of urea, a colorless, odorless, highly soluble crystalline solid. At higher concentrations, urea can be used in the treatment of many skin conditions like dermatitis and psoriasis. At lower concentrations, it is excellent for plants, which use its nitrogen to divert the sun's electromagnetic energy into chemical form. Urea also, when hydrolyzed by bacterial urease, turns into ammonia and carbon dioxide, giving old urine its pungency. Along with urea, there are also small amounts of chloride, sodium, sulfate, and potassium in urine, as well as an array of hormones and metabolic wastes that vary with diet. Notable in this mixture is urobilin—a chemical only recently discovered to derive from the decay of red blood cells—which gives urine its most distinguishing feature, its pale-yellow hue.

The evaluation of urine was one of the earlier forms of medical practice. A clay tablet at the British Museum records the Sumerian terms for the varieties of urine, which they knew as penis water 𒃼𒀀 𒇷 𒃼𒀀 𒈗 —from *sinatu zalmi*, dark penis water, to *urpati sinatu,* cloudy penis water—and an Assyrian tablet offers diagnoses for those with urine like beer dregs or beet juice or paint. What we now think of as laboratory medicine—the biopsy of a mole or the DNA sequencing of the viruses in one's mucus—began with little more than a vessel and some urine. When you inspect the contents of your toilet bowl, you are reenacting one of the oldest known medical analyses.

Urinalysis still provides invaluable information, and the mere appearance of urine can tell you a great deal about the state of your being, though there's a lot of misinformation out there. Despite what Verbal Kint maintains in *The Usual Suspects*, for instance, dehydration can't cause your urine to "come out like snot."* In general, it's best to think of urine in terms of those color charts at the paint store where each hue is presented in different tints, tones, and shades. The color of urine acts as a clue to potential underlying conditions within your body,

---

* Cloudy urine or snot-like discharge, especially if foul-smelling, is more likely an indicator of an infection and is often associated with sexually transmitted diseases like gonorrhea and chlamydia, though Kaiser Söze probably knew that.

while the depth of the shade indicates the severity. Healthy urine is pale yellow, for example. If your urine is very pale or even clear, you've been drinking too much water, putting you in danger of developing hyponatremia, a potentially fatal sodium imbalance,* while deeper yellows and browns indicate increasingly severe levels of dehydration. Other hues can point to a range of issues, with pinkish-orangish urine indicating blood (enlarged prostate, bladder cancer, kidney stones) or possibly that you've been eating beets. Bright yellow means too many B vitamins, and green urine is caused often either by certain antidepressants or by blue food coloring. Purple is likely porphyria (if you want to be sure, put the urine under a blacklight and see if it glows pink).

Though the subject has not been explored in much depth, there's also a lot to be learned from the aroma of urine. It isn't only asparagus that gives a distinctive smell to the fluid, after all. Many substances—from coffee and tea to fennel, garlic, and strawberries—endow urine with their own unique aromatic compounds, and these scents carry with them important information about the body. Urine that smells of fish is a sign of the metabolic disorder trimethylaminuria, and urine that smells of ammonia is a sign of an infection in the urinary tract. Certain cancers give urine a distinctive scent that, while undetectable to the human nose, can be identified by dogs trained

---

* Perhaps the most misrepresented scientific research relates to water consumption. While the human body requires roughly eight cups of water each day to function, much of that is derived from the foods we eat.

to smell human urine. Methoxyphenols, which have smoky or caramel hints, can indicate woodsmoke exposure of those near forest fires, and the inclusion of vanillin in pharmaceuticals can help doctors assess whether patients are abiding by their treatment plans. One of the first diagnoses by scent was for a once-rare disease that caused the extremities to wither and die. The disease was believed by Aretaeus to be "a melting down of flesh and limbs into urine." Patients suffered from an insatiable thirst, but no sooner had they drunk than they felt the urge to urinate. It was as if the urine was using the human body like a ladder, he noted. It was as if a human being could be subjugated by its own water. We recognize the disease now as diabetes, and the diagnosis can still be made as it was in Aretaeus's time, by the sweetness of the urine's smell.

Golden Ages

On an island off the coast of Sweden, human urine is now diverted from solid waste, dehydrated, and sold as a commercial fertilizer, and in Singapore, close to a billion liters of wastewater are captured and reused daily. Israel repurposes 90 percent of its sewage for crop irrigation, while several jurisdictions in Texas and California now capture their liquid waste, treating and recycling it into potable water. Indeed, the saving of human urine is now so commonplace that it's hard to say when exactly human beings first began to keep their own urine. While it's entirely possible that the earliest humans inspected their urine for clues to the function of their own organism, genetic analyses of migratory patterns depict a prehistoric population in near-continuous flux, and given this restless, wandering existence, it seems most likely their urine, if kept, was saved only briefly, on a case-by-case basis. Only with the Neolithic Revolution some twelve thousand years ago, and the ensuing shift to relatively stationary lifestyles, would the reservation of urine even have been given the opportunity to blossom into a concerted, culturally significant activity. And yet it seems little urine was kept.

The Cucuteni—maybe the first humans to live in cities—might have saved their urine, but in excavations of their dwellings along the lower Danube no archaeologist has discovered any evidence of the practice. And any hypothesis that the culture kept its urine seems untenable, given scholars' suspicion that every sixty to eighty years the Cucuteni would ritually burn their own settlements to the ground. We can say safely that for the great pastoral nomads of history—the Scythians, the Xiongnu, the Huns—urine-saving in any substantial degree was simply an impossibility. We know from medical texts vitrified in the burning of Ashurbanipal's library that the people of ancient Mesopotamia were aware of urine—what it meant when it could not be made, when it became clouded and dark, when it burned. But to judge from these texts and from the culture's primitive sewers, it seems that with whatever admixture of feelings they regarded the substance, they were mostly terrified of it, distrusting it so fully that they established in their pantheon a place for a lavatory demon, Šulak. And even though it was relatively common for civilians and royalty during the Han dynasty to be entombed with a toilet, when considered, this seems to indicate not a desire to save urine but a presentiment that even into the afterlife it would pursue our hapless souls.

Perhaps because of its role in revealing the body's inner turmoils or because of Vespasian's famous dictum—*pecunia non olet,* "money doesn't stink"—urine has long been suspected of transformative properties. Just as some today believe that the venom in a jellyfish's sting can be neutralized by urine or that two hundred milliliters of urine, imbibed once daily, will unburden the depressed mind, Sir Thomas Browne reported that physicians of his day convinced patients that "there is the book of fate, or the power of Aarons brest-plate in Urines." And Sung Tz'u, in his 1247 forensics manual, *The Washing Away of Wrongs,* gave these instructions for reviving the slain:

> Take some frankincense and myrrh and a large soap bean, all ground until they have disintegrated, one-half cup of urine, and one-half cup of good wine, and boil these together. Have the victim swallow the mixture. After that, take dolomite powder, some cuttle-fish bones, or dragon's bones, powder them and cover the mouth of the wound.

Bernardino de Sahagún, in the *General History of the Things of New Spain,* wrote that fresh, warm urine was commonly used by the Aztecs to cleanse head wounds or to alleviate certain internal maladies, but neither in Sahagún's history of things "both good and evil" nor in the texts of Sung Tz'u or Browne do we find any mention of the actual keeping or collecting of human urine.

In fact, before the construction of the world's first wastewater recycling plant in Windhoek, Namibia, in the late 1960s, the best example of large-scale, long-term human urine conservation took place in ancient Rome. Even when the city was still in its urine-saving infancy, before Agrippa punted earnestly down the Cloaca Maxima, many of Rome's back alleys were simply troughs where the poor made their water and where desperate mothers took their newborns, hoping the fumes might relieve both mother and child of the burden of a new life. By the time of Nero, the saving of urine had become an institution in the Roman Empire. In the vast apartment blocks where much of Rome lived, people urinated into small pots which, when full, were emptied into much larger vessels beneath the apartment buildings' stairs. When away from home, a Roman could pay to use a public latrine and relieve himself among the figures of gods, or for no charge at all, he could pause at a street corner and use the vessels there, provided by the guild of fullers. From these vessels, as well as from the latrines, the fullers of the empire—human urine conservationists perhaps unrivaled in all of world history—acquired the materials necessary for their work. They gathered Rome's urine into large stone basins, and in these basins men and boys performed what Seneca called the *saltus fullonicas*, the fullers' dance, using their feet to trample clean the city's woolen garments. As evidence of the degree to which this system of saving urine had become an integral part of Roman life, we know that Nero, long before his flight from Rome, long before he is said to have recited Homer and cut his own throat, levied his infamous *vectigal urinae*, the urine tax.

But then with the slow, heaving collapse of the Roman Empire, urine almost wholly ceased to be collected in any meaningful way. It wasn't until 1669 that a man by the name of Hennig Brandt, having married a wealthy widow, began to save his urine. Whether this urine was kept in pails or barrels or empty wine casks we cannot know, but scholars suspect that the urine may eventually have totaled in the thousands of liters. Brandt, due either to an obscure belief or simply to necessity, allowed the urine to putrefy in his family's dwelling in the Michaelisplatz area of Hamburg. This coastal city would become a formidable shipyard in the centuries to come and would be the very epicenter of the Allied firebomb raids of 1943, in which tens of thousands would burn, but at the time that Brandt began to boil down his store of urine it was only a place of small wooden structures where salt and cod were traded. He expected, of course, to make gold or, failing that, a substance capable of turning all others into gold. He made neither. Instead, when all the urine had been boiled away and when the precipitate too had burned until it smoked, he was left with a fine white substance that, in the absence of lamplight, glowed an unnatural blue-green.

He had made one of the great chemical discoveries of history, seeing for the first time with human eyes the isolated element of phosphorus, the element that is perhaps most essential to the continuation of life, a component in every plant and animal, as well as in the bombs that would eventually transform Hamburg into an inferno, but despite this revelation, Brandt, unable to make phosphorus become gold, eventually abandoned his urine experiments. Maybe this is not surprising, maybe given the history of urine, we cannot expect Brandt to have appreciated exactly what he discovered.

After all, in the middle of July in 64 AD, the Great Fire burned Rome for two full days. Had the fire been hot enough, lasted long enough, it could have evaporated the urine in the fullers' basins and burned the precipitate into phosphorus. At night, had they looked, the Romans might have seen the element shining. It might have seemed a sign. Had they touched it, had they felt it burn and burrow into their skin, as it was rumored to have done to those present at Hamburg's destruction, riddling and scorching their bodies with an elemental heat, so that even upon plunging into Hamburg's canals still their suffering went unrelieved and they begged to be shot, had the Romans felt this, it might have seemed to prove all that had been feared of urine, to show that the materials of our composition are, finally, only as comprehensible as they are painful. But if phosphorus was created in the Great Fire, its faint light went unnoticed. As Tacitus reminds us, in the days following the fire, there was no darkness in Rome. The inhabitants of that greatest of cities were busy with the work of retribution. Those thought to be responsible were covered with the skins of beasts and beaten, they were taunted, torn by dogs, and nailed to wooden uprights, and they were burned, when daylight had expired, to serve as illumination for their own trials.

III

Blood

> So the blood becomes bewildered
> living becoming
> nothing but tenacious pageantry
>
> Will Alexander, *The Sri Lankan Loxodrome*

Terms

n. 1. lifeblood, vital force, claret, cruor, juice

n. 2. line, lineage, roots, stock, heritage, extraction, descent, pedigree, race

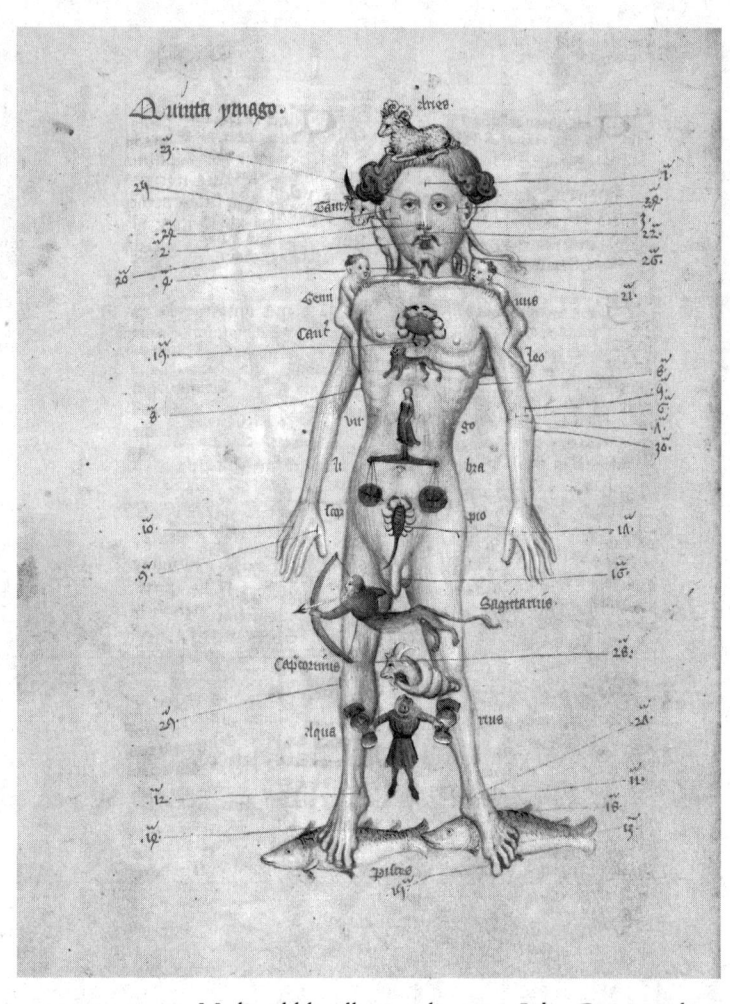

Medieval bloodletting diagram, *Liber Cosmographiae*, John De Foxton, 1408

Biological Prologue:
DSM-5 BII Caveat Emptor

My first serious girlfriend was a tall and moderately rugged catcher for the high school softball team. Her name was Jessica Suarez, and one summer afternoon, we set off together on a stroll through town. We walked, and we talked, and eventually—it was probably her initiative—we began to hold hands. We were in it deep at the time, as anyone must be who gets a rush from holding hands. We were silent sometimes while we walked, silent and smug.

I liked Jessica a lot. She'd grown up on the edge of town, in a house that backed up on the forest. She knew how to split wood, plow snow, drive stick, paint a watercolor of nasturtiums on a windowsill. She could sling an errant foul to second with enough pace it left a lump on your palm. She was one of the toughest and most competent people I knew, and sometimes she put my hand on her breast. I loved her.

We were on a stroll this day. It was late spring, the robins were doing it in the trees, the trees were doing it, via the bees. The world was alive, and it was great. Then we came around the corner, and there was this little church with one of those plastic lawn signs out front with a squat red cross.

"Hey," I said. "Want to give blood?"

I don't know why I said it. Jess immediately shot me down. She didn't say no, she just gave a quick shake of her head and pulled me along. We were passing the church then with its big set of double doors propped open. Two dour elderly women peered out from the doorway. We couldn't see into the darkened interior where the blood drive was going on, but as we passed, some of that crypt-y church air lapped around our ankles. It smelled like mildewed hymnals and flaking lacquer and eau de toilette and maybe just faintly of the rubbing alcohol they use to wipe down veins. As we waded through that pool of cold air, Jessica shimmied, then she tilted her head toward the church and leaned in that direction. She'd changed her mind, I thought. We were going to give blood after all. She was so spontaneous!

Then her knees buckled, and she fell unconscious to the lawn.

We'd been moving along at a decent clip, and her collapse happened so suddenly that my momentum continued to carry me forward for a second. But with her catcher's hand somehow still clamped viselike onto my own, my movement was arrested, her mass rubber-banded me backward. And then I stood there, still gripped by the hand, precariously balanced on the tips of my toes, leaning over her body, which was just kind of lolling aimlessly on the ground. She seemed to be trying to get up. "Flarbleclampn," she said. "Mlarbnflup." I looked up at the old church women, the old church women looked at me, I looked at Jess, they looked at me looking at Jess, and that's how I first learned about BII.

BII (blood-injection-injury) phobia is in the *DSM-5* alongside spiders, lightning, enclosures, heights, crowds, journeys alone, drowning, suffocating, and all the other classic Hitchcockian phobias. But BII is different. Like the others, it's characterized initially by the standard autonomic fight-or-flight response. The body undergoes tachycardia, a rapid and irregular beating of the heart, as well as a jump in blood pressure. But in BII phobia this initial reaction abruptly reverses. In a moment, the heart rate drops dramatically, blood pressure decreases, and the rising panic melts into a woozy vasovagal response: sweating, nausea, ringing in the ears, narrowing of the field of vision to a smaller and smaller circumference, and you faint. Nobody knows why BII exists—it doesn't seem like it would confer much evolutionary advantage if the sabertooth bit you and you passed out—but exist it does.

Those who suffer from BII show "prioritized visual processing of blood." The sight of blood seeping from a fingerstick, pooling in a blood bag, or spattered down a roadway, all of these can bring a BII phobic to their knees. They have difficulty looking at red spots. Some, like Jessica, may faint at nothing more than the thought that someone, somewhere nearby, is about to give blood (one of the common persistent reveries of BII phobics is the image of a needle breaking off beneath the flesh). They do not want to know that the human body circulates twenty or thirty thousand liters of blood every day. They would rather not learn that Seneca's circulation was so poor in his old age that when he tried to cut his wrists, nothing came out and only when he took a hot bath did the blood begin to flow. They don't want to listen to the hundred thousand beats of the heart in a given day or know that the aorta is roughly the same diameter as their garden hose. There are reports, among BII phobics, of ketchup-induced nausea.

It is to Jessica Suarez, wherever she is now, and to the estimated several million sufferers of BII phobia in the United States that I would like to make a suggestion: set down the mug of hot coffee; get away from the basement stairs; cease operating heavy machinery; find a comfortable chair in a location away from sharp objects. This chapter contains a great deal of blood.

Having said that, for the percentage of BII sufferers who find the pronunciation of the word *blood* itself excruciating, the vowel rising up from the throat and filling the mouth with pressure before parting the vermilion lips and washing out across the tongue—"Blooooood"—for those good folks, I come bearing potentially uplifting etymological news. While the word *blood*, in one form or another, is inescapable in many European, Middle European, and Old European languages [old Frisian *blōd*, Old Dutch *bluot*, Middle Dutch *bloet*, Dutch *bloed*, Old Saxon *blōd*, Middle Low German *blōt* or *blōd-*, Old High German *bluot*, Middle High German *bluot* or *pluot*, German *Blut* or *Blüt* (as in the *Deutschblütigkeitserklärung*), Old Icelandic *blóð*, Old Swedish *bloð*, Swedish *blod*, Old Danish *blod*, Danish *blod*, Gothic *blōþ*, and Crimean Gothic *plut*], there is no clear proto-Indo-European root. There's a mystery here. Words don't just get invented. They evolve out of older words. But start digging back through the etymological record, and *blood* just suddenly disappears. It's like it came out of nowhere. And maybe it did. As some etymologists now suggest, it's possible that prior versions of *blood* don't exist for one simple reason: *blood* is only a placeholder for a much earlier term, a word that, for fear of incurring some supernatural wrath, was never meant to be spoken.

The Price of Blood

The blood center in Santa Fe, part of the Vitalant network, consists of two storefronts in a mostly vacant adobe strip mall in the foothills of New Mexico's Sangre de Cristo mountains. There are other businesses in the vicinity—a martial arts studio, a frozen yogurt chain, and a wellness center called dancingbones—but it isn't the sort of place you stumble upon accidentally. It's at the intersection of a couple of busy roads but set back from the street in such a way that you'd never know it was there if you didn't go looking for it. As if compensating for this lack of road visibility, the windows and interior are plastered with signage. "Make a difference!" "Type O? Consider a whole blood donation!" "Sign up for Hero Rewards! Give the *gift* of life, get a Cold Stone Creamery *gift* certificate. Only 700 Hero points."

They like me at the blood center. I'm youngish and healthy-ish and my paperwork is in order. I'm O+, the least remarkable of blood, but I bleed fast, and at a little over six feet, I qualify for maximum donation volumes. They'll drain a liter of plasma off me in under an hour and a half. But the main reason they like me is my veins. The men in my family, through some genetic lottery, all have pronounced venation in their arms. I roll up my sleeves, and the phlebotomist sucks in her breath in a reverse whistle.

"Gorgeous."

There's something actively eager, even gleeful, in her tone. In a fairly steady stream of talk, she mentions the veins no fewer than seven times, the exclamations acting as a sort of punctuation for her life story.

"I lived all over—Denver, Seattle, Oregon, Fort Worth—but after the first divorce, with two kids—roll up those sleeves for me again, awesome, just awesome—I decided it was time to come back to New Mexico. Well, the two boys are in college—we'll just put this tourniquet on there, even though you don't really need it—but our surprise, she's eleven, and of course she's got some dyslexia."

She yanks the band tight.

"And autism."

She yanks again.

"So there you have it."

Blue ridges strain against the skin in the crook of the elbow. "I guess I've got my pick of the litter," she says and runs a finger over the vessels. Then she rips open one of those little packets of the sort they hand out with BBQ ribs and swabs the skin with a scratchy alcohol wipe.

"Eenie, meenie, minie, moe . . . I'm going for this bad boy."

We both look down at the bulging vein.

"Mmm." She meditates over the prospect. She's wearing a silver charm bracelet and a lanyard printed with Pokémon characters.

"I could stick people like you all day long," she says. Then she drives the needle home.

When I donated in college, the drives were held in the

university's deconsecrated chapel. In place of pews, beds had been spaced at intervals across the open room. The bleeders lay in row after row, their heads slightly elevated, as if struck en masse by some irresistible fatigue. A saline drip was suspended above each while a bag below, marked with inscrutable codes describing the donor's specific antigens, collected the blood in vacuum silence. The already-bled lay there too, disconnected from their apparatuses, ripping open packages of Fig Newtons or Oreos sized specifically for the caloric needs of the Red Cross and chewing them still half-reclined as their eyes wandered over the ceiling. The phlebotomists, meanwhile, drifted from one patient to the next, operating out of pure rote memory, disconnecting and reconnecting the various tubes, looking up to a clock on the wall as they counted the pulse, rubbing alcohol on the inside of the elbow and tying off the arm with a tourniquet of latex, chatting thoughtlessly or in seeming thoughtlessness as they eased the needle in. When your name was called, you took your place among them and rolled up the sleeve.

The high-vaulted ceiling, the stained glass panels depicting the stations of the cross, the gargantuan and pensive organ pipes, and the busy ministering of phlebotomists—you felt yourself being drawn into this scene, as into some obscure and ancient ritual. A hidden side door opened from the paneling, and a man appeared with a hand truck and began loading up the coolers of collected blood, and at any moment, it seemed another unseen door might open, producing a train of elders in vestments, swinging a censer and chanting the genetic sequences of the ABO blood group.

At my new blood center, by comparison, the scene is mundane. On one wall hangs an enormous television, and the beds are all angled to provide maximal viewing. It's October, which means there's a horror movie marathon in progress, but since it's the middle of the afternoon, there's nothing too grisly on yet. I catch the very end of the neutered transhuman love affair between Casper the Ghost and Christina Ricci, then the macabre antics of the Addams family start up. The only gore comes when Pugsley and Wednesday Addams (Ricci, again) reenact the final duel in *Hamlet* for their talent show. A strike to Wednesday's left wrist causes a little jet of blood to spurt across the stage.

"'A hit,'" says Pugsley, "'a very palpable hit.'"

Wednesday stares down at her hand with a remarkably convincing look of surprise, then turns and severs his left arm at the elbow, unleashing a hydrant of blood. He takes a final swing, slicing open her throat to geyser-like effect, and she spins dramatically toward the audience, eyes rolled back, soaking the first few rows in blood.

"'O proud Death,'" she recites. "'What feast is toward in thine eternal cell...'"

It's a great speech, this final soliloquy, but when I look around, I seem to be the only person watching. Everyone else has their eyes closed or is on their phones.

There are all kinds of folks here. There's an old man down from Tesuque Pueblo for the day with an hour to kill, a young dude in one of those short-brimmed cyclist's hats, a lanky Asian woman with a near-empty blood bag at her side who seems to have been here for hours because, as one of the phlebotomists tells her, patting her on the arm, "Some days you're the turtle, hon, and some days you're the hare, and today you're not the hare."

Back in the old days, the "blood-on-the-hoof" era of transfusion history, we would have gotten the call to donate when someone with our blood type was hit by a streetcar or suffered postpartum hemorrhage, and we would have hopped on a horse and hustled to the hospital as quickly as we could. Our veins would have been spliced and sewn onto the veins of the recipient via anastomosis, and we wouldn't have been watching Christina Ricci pretend to pretend to die on TV. We would have been watching another human being physically return to life. But today a whole blood donation no longer requires that the donor be physically connected to the recipient. We don't lie down next to a recently trampled stable boy and watch as our blood ushers his world back into view, but there's still this sense, even if it's not immediately palpable, that something weird and vital is occurring in this room.

What brings us all here, after all? *What's it worth to us?* Sure, there are the Hero points that Vitalant doles out for successful donations. You can redeem them online for Hero merchandise, which is pretty sweet if you've really been hankering for a phone charger or a Hero T-shirt. But we're not in it for the points; we're brought together by something else, something that hovers above the usual earthly concerns. We're saving lives. That's the big idea, I think. And it doesn't matter if it's the mother of triplets, the child with cancer, or the hemophiliac linebacker. It doesn't matter if the person is nice, or signals before changing lanes, or if you agree with their views on fracking, abortion, and pentobarbital. It's deeper than that at the blood center. Not always but sometimes, as a blood donor, you feel you're surfing a more generous societal frequency. The ideals of socialism—from each according to their ability, to each according to their need—seem to be alive and well. And it feels good, really good, in a way that's not merely light-headedness, to be freed temporarily from a world where it seems like everything has a market value. It feels good to know there's something priceless in your veins and you can give it away anytime you like, so long as you wait at least eight weeks between visits. That's what has brought me back every eight weeks for much of my life.

And it's a little demoralizing then to discover what happens when you walk out the door of the blood center. They turn around and mark up your donated blood and they sell it, usually for about two hundred dollars a pint. They might sell it to pharmaceutical companies or academic research institutes, but more often they sell it to an organization that oversees blood distribution for a certain geographical territory, and usually it ends up in the hands of one hospital or another. And then the hospital turns around, adds its own markup, and throws in the costs of transfusion, and suddenly what you gave away for free ends up costing that kid with cancer a few thousand dollars (or rather that kid's parents' employer's insurance plan), and the miraculous red stuff you thought was a big altruistic middle finger in capitalism's face turns out to have been just another commodity on another market in a world where everything gets bought and sold eventually.

THOUGH IT'S NOT TOO CRAZY WHEN YOU THINK ABOUT it. There are some fixed costs associated with collecting, testing, storing, and delivering fresh living blood to those who need it. You need office space and EldonCards and apheresis machines, and you need refrigerators, and you need emergency generators for those refrigerators. You need phlebotomists and lab techs, of course, but you also need website developers and custodians and someone to answer the telephone and turn off the lights at the end of the day. And then there are the tests. You need to make sure the blood doesn't have hepatitis C or HIV or COVID-19, and you need to make sure you're not going to accidentally inject B+ blood into an A– person. When you take that total and divide it by the number of pints collected, a few thousand dollars doesn't seem so crazy.

The funny thing is that while the blood market does exist, it's hardly a market at all in the conventional sense. Largely because the blood banking system came of age in the United States during the nationalistic fervor of World War II, a certain amount of altruism is built into the model. Unlike most of what goes on in hospitals, with blood there's an implicit assumption that people will give it away and blood banks will sell it without profit and hospitals will resell it without profit, and this means the whole system operates without market pricing, which is sort of the social democratic ideal everyone in my blood center wants, right? They might not like that a woman who needs ten pints after a difficult birth ends up shelling out the equivalent of a down payment for a modest home, but at least nobody's trying to make a buck off their grand karmic gesture. Blood isn't an Uber. It's the very stuff of life. All of us donating would be more than a little pissed if someone got surge-priced for our blood because he got in a car accident on Christmas.

But this also means, according to blood economists Robert Slonim, Carmen Wang, and Ellen Garbarino, that while the market is still "heavily influenced by standard economic forces including supply and demand, economies of scale, and moral hazard," it can't always adapt fluidly to these pressures. Because the price of blood is basically fixed, the market has particular trouble matching aggregate supply to demand. Which is to say, the price of whole blood doesn't go down if there's a glut of donors, and it doesn't rise during the seasonal winter shortage. For obvious reasons, that's not really ideal. Like those astronauts who can't stand up when they return to Earth because their vasculature no longer adjusts to gravity, the blood market can't always get the blood where and when it needs it.

But even if you wanted to figure out the price of a pint of blood, it isn't just a matter of calculating supply versus demand. Though the very miracle of transfusion is that the blood of one person somehow works in another person's body, pints aren't just interchangeable. If you're trying to figure out what a given pint is worth, you've first got to take into account two inherent attributes of blood: antibodies and antigens.

Most people are probably familiar in a general way with antibodies. The body reacts to an infection by making these little Y-shaped proteins and releasing them into the bloodstream. They circulate, floating around in the plasma, but when they brush up against the specific bacterium or virus responsible for that initial infection, they latch on, either disabling the foreign body or marking it for removal by other components of the immune system. (This is how vaccines work; they prompt the body to make specific antibodies without actually making you sick.) Antibodies aren't yet a huge determinant of the value of a pint—we're still only coming to understand the ways they might be used in fighting disease—but the COVID crisis showed just how valuable they are on premise alone. On the plasma market, which for weird historical reasons is a highly flexible international for-profit market, the convalescent plasma of COVID survivors retailed for as much as eight times its prepandemic price.

Currently, however, the more fundamental value of blood derives from the antigens on red blood cells. Your red blood cells are shaped sort of like chubby contact lenses, and affixed to the outside of each cell, like tiny molecular tags, are countless proteins and sugars. These are the antigens. When you say your blood type is A, for instance, what you're really saying is that the surface of each of your red blood cells has an A sugar antigen on it. And when you say your A blood is +, what you're saying is that of all the proteins in the Rhesus (Rh) system of antigens (from C to c to E to e to S, s, U, LW, Fy5, weak D, and partial D . . .), you have the RhD protein on the outside of your cell.

It's not yet clear exactly what these antigens do, though we presume each has or once had a purpose. There are hints. Type As get more stomach cancers while Type Os get more ulcers, whereas Os get more bubonic plague and As get more smallpox. ABs might be slightly more resistant to malaria, and O seems to offer resistance to SARS-CoV-2 infection. We know from coalescence analysis that A was the first blood type. The O mutation occurred about 3.5 million years ago, and B entered the picture about a million years ago, which is to say several hundred thousand years before *Homo sapiens* existed. AB is just the chance combo of a parent with A and another with B.

It's easy to get lost here. O, for instance, was previously named 0 to signify a lack of either A or B antigens, and before that it was known as C, while in Russia, where they still use the Polish blood nomenclature of Roman numerals, it's known as Type III, which some old Soviet veterans still have tattooed on their shoulders for ease of wartime transfusion. For the purposes of determining the value of a pint of blood, your various antigens, or lack thereof, help determine whose blood your immune system will accept and who can accept your blood. If you get a transfusion with different antigens, the antibodies in your blood will recognize the transfused cells as enemies, and your blood will begin to attack itself. The chest aches, the veins burn, the urine turns a ruddy brown. The blood literally coagulates in the veins. It doesn't take much to kill the recipient: just a tablespoon or two. Some people put more cream in their coffee.

It's that gumming up that makes certain types of blood innately more valuable than others. O–, for instance, is a workhorse. It has neither AB nor RhD antigens, which means you can stick it in just about anybody regardless of their ABO and Rhesus antigens and be confident the blood won't agglutinate in their veins. Because you don't need to worry about the blood type of the recipient, it's the break-in-case-of-emergency blood. When a gurney gets wheeled into the ER, the doctor reaches for a bag of O–. AB+, on the other hand, which has all three antigens, is perfectly wonderful blood, but it's only fit for the select 5 percent or so of the world population who are themselves AB+, making it of limited utility in certain circumstances—if you were transfusing among the Blackfoot of North Dakota, for instance, a majority A population.

And the ABO and Rh antigen systems are just the biggies. There are millions of antigens on the surface of red blood cells. You can be Duffy negative, for instance, or Diego positive. If you want to go full law firm, you can be Kell, Kidd, Cellano, and Wright. These then go all the way down to the absolutely rarest types of blood, until you get to Rh_{null}, a blood so scarce that fewer than fifty people are known to have it in their veins. These rarefied types, the Duffy negatives and Rh_{nulls}, aren't useful much of the time, but when they are required, nothing else will do. Thickets of customs officers are braved and helicopters commandeered and whole surgeries relocated all to bring together a single pint of blood with the one person who can use it. In fact, because this blood is so precious and because it is often subject to many of the same trade sanctions as other goods, it is often simpler to actually transport an Rh_{null} donor to the recipient rather than risk having a pint languish at the border.

And this gets down to the other factor that makes pricing blood different from pricing, say, steel or microchips. While the logistical challenges of navigating time and distance are part of every market, the value of blood is acutely determined by one inescapable feature of the blood itself: it goes bad.

Since blood has a shelf life—forty-two days if kept at 4 degrees Celsius—its real-world value will always be a function of when and where it is taken. There are ways around this dilemma. You can divide the blood into components, centrifuging out the red blood cells ($200) and the plasma ($60) and the platelets ($500). The platelets go bad in a week, but the red blood cells can be kept frozen for years. (The armed forces keep tens of thousands of units of frozen red blood cells floating on ships in the Pacific and Mediterranean theaters at all times, to be used in the event of a sudden large-scale conflict.) The plasma can be dehydrated and stored pretty much indefinitely. But if the blood is kept as a complete pint—whole blood, as it's called—the clock starts ticking. This is why a generic pint has more value close to the urban centers, where much of the population lives, than it does in remote rural areas. A pint of blood in New York City is more likely to be used than a pint in North Dakota. And the more esoteric the blood, the more its value is location-dependent. Donate Duffy negative in Canada when it's needed in the United Arab Emirates, and it will be rotten by the time it makes its way through customs. It might as well be a bag of O+.

Which all makes you start to think Slonim, Wang, and Garbarino are on to something. The market is inefficient. And there are moments, even just when you're considering the humdrum domestic distribution of blood, when the peculiar contours of blood donation reveal how bizarrely blind the market is. There were times during the COVID pandemic, for instance, when you couldn't schedule a routine surgery because it wasn't clear whether there would be any blood when the date came around.

Strangely, though, the moments that most make you wish for pricing are those when there's *too much blood* on the market. Not too often, but more often than one would like, the market is glutted with donations. After mass shootings, for instance, like the one at Sandy Hook Elementary School in 2012 or the one at the Route 91 Harvest Music Festival in Las Vegas in 2017 or at Pulse Nightclub in Orlando in 2016 or Robb Elementary in Uvalde in 2022 or the El Paso Walmart in 2019 or the Tree of Life Synagogue in Pittsburgh in 2018, many people feel an understandable urge to help in some way, which is laudable, so they line up at OneBlood in Orlando or at the Pittsburgh Penguins Stronger Than Hate Blood Drive, and they give blood, a great deal of blood often—28,000 pints after the Pulse shootings—even if there aren't actually many survivors to use it.

This peculiar market behavior was most notable on September 11, 2001, and in the days following. In Gainesville, in Tulsa, in San Diego, in cities and towns across the country, donors flocked to blood centers, some already en route before the second plane hit. They sat on floors, looking up at the television and waiting for their names to be called. They stood in lines that snaked down the block and around the corner. And as images of those blood centers themselves became newsworthy, it created a positive feedback loop. Newscasters, encouraged by the Red Cross, asked viewers to donate, viewers did so in droves, and as more and more people began to feel blood donation was the correct ethical response, a herd mentality took hold. Robin Williams donated blood, as did Yasser Arafat. On September 13, when the nation's blood supply was already at record levels, C-SPAN aired footage of members of Congress reclining side by side in the Cannon House Office Building for a secure congressional blood drive. It was, according to Representative Deborah Pryce of Ohio, a "bipartisan effort."

So was the hearing a year later by the Subcommittee on Oversight and Investigations, at which various parties involved in the collection of blood during that period testified that in the days following September 11, the nation processed an estimated 570,000 additional pints, a remarkable achievement, both morally and logistically, representing the country's generosity of spirit, its tireless work ethic, and tens of millions of dollars in blood bags, tubing, storage, and overtime. Remarkable too when one realized that of the roughly half a million pints donated to help the victims of September 11, fewer than 260 had been used for that purpose while about 200,000 had been quietly and systematically destroyed.

THE BLOOD MARKET AS WE KNOW IT TODAY CAME OF age largely over the four years spanning the end of the Great Depression and the beginning of World War II, 1937 to 1941.

In the 1930s, blood was not usually saved. Instead, hospitals kept paid donor registries. If a pint of A− was required, they went down the list of A− donors, calling each individually until they found someone who was willing and able to donate. The donor came to the hospital, had their blood drawn, got paid, and the blood was already being transfused by the time they got home. It was a one-to-one transaction, in other words. Although it was facilitated by the hospital, for all intents and purposes it was a private sale.

The limitations of the model are obvious. With a patient lying on the operating table, there was hardly time to call through a list of numbers hoping to catch someone at home. And it was the Great Depression. Even when the right blood could be found, at Cook County Hospital, a public facility where Bernard Fantus oversaw the blood program, many patients couldn't afford it.

Inspired by a Soviet surgeon's account of draining and refrigerating for later use the blood of a man who had died in a car crash, Fantus began to conceptualize a better approach. To eliminate the frantic calls to the donor list, he needed to have enough blood on hand at all times so that he could see to the needs of any patient. The question was not how to store the blood. He was able to keep it for up to ten days at the time. The question was how to get and continually replenish such a large supply. He found an answer in the realization that while many of his patients lacked money, they had blood. If there were a system for crediting and debiting donations, he could ensure blood continued coming in as it went out. So he created one. Each patient had their own blood account. They could make deposits to this account, or they could make withdrawals. All pints were valued equally on a blood account. Just as a dollar of corn is worth the same amount as a dollar of whiskey, a pint of A+ was worth the same as one of O–. By making pints interchangeable, he'd turned them into a kind of currency. He called the system a blood bank.

An efficient setup, it was quickly adopted by hospitals across the country, and two years later, as Britain became embroiled in World War II, the United States used the blood bank model to create its first national blood collection program, Blood for Britain. The goal was to collect the most blood plasma in the safest manner in the shortest time, and the program, based in New York, was placed under the direction of a Black surgeon named Charles Richard Drew. Drew's problem was both logistical and philosophical. He needed to expand and systematize Fantus's concept while abandoning its central pillar—the debits and credits that provided an incentive to donate—and he needed to do so quickly.

It would have been hard to find someone as dedicated to the work as Drew. During his time in New York, he lived away from his family in a small apartment overlooking the Hudson River, and it seems he did little but think about blood and send letters home. "I am working all day (and a good part of the night) in the laboratory making determinations of the components of blood," he wrote to his mother. "This morning at 1 a.m. I was called in to interpret the blood findings on a very sick patient." His letters to his wife were more meditative in nature:

> *I know you think this is all just a little bit heroic, sometimes I do too, and laugh at myself. But most of the time I am deadly serious about it, as though I had nothing to do with it but simply carry out commands given me by some inner force which never wants to play.*

The name of his daughter goes a long way toward explaining how possessed he was by his work with blood. He named her Bebe, after Blood Bank.

His diligence bore results. In a little under six months, he had created a system in New York City whereby donors were solicited and blood drawn, tested, typed, separated, refrigerated, shipped, and tracked, all under sterile conditions, collecting 14,556 donations for the British cause. In fact, he was so efficient in his work that in 1941, as it became clear the United States would likely join the war effort, he was asked to help pilot a national blood collection effort begun under the auspices of the American Red Cross. The national system of blood collection was still in the process of expanding across the nation on December 7 of that year, when Pearl Harbor was attacked.

This, in many ways, was the moment when the blood market as we now know it—a nationwide system of volunteer donors banking blood for unknown recipients—came into being. Within three months of the attack, weekly donations had increased twelvefold.

"If I could reach all America," said General Eisenhower at the time, "there is one thing I would like to do—thank them for blood plasma and whole blood."

AT THE BLOOD CENTER, I'M STILL THE ONLY ONE EN-joying the Addams family. During commercial breaks, they've started advertising the next film to be screened, *Interview with a Vampire*. There's a chance I'll be able to catch a little of the flick. I'm not here for a regular donation. I've signed on to do apheresis, which takes a while. It's basically a centrifuge and pump setup. The blood gets drained into a spool of tubing, cooled, and spun until it separates into its component parts: plasma, platelets, red blood cells. One component gets siphoned off, then the whole flow is reversed, and the rest of the cooled blood gets pumped back into the donor. The procedure repeats—a few minutes out, a few minutes in—until the bag is full. It's not speedy. It can take a couple of hours, but I like it because it's more efficient. By drawing off only one part of the blood, whether the plasma, red blood cells, or platelets, it means they can take a whole bunch of that specific component. If there's a need for platelets, for instance, an apheresis platelet donation can provide as much as six whole blood donations.

But it's a strange sensation being hooked up to an apheresis machine. It takes time for the phlebotomist to sort out all the tubes, and there's this special cylindrical bag that gets wrapped around the centrifuge. There seems to be an art to arranging it that my phlebotomist has yet to master. She routes and reroutes the tubing three times, but she can't satisfy herself that it's quite right.

"Now and then this tube gets caught," she explains. "And then when it starts to spin . . ." She trails off. "It's not pretty."

"What happens?"

"Oh." She's matter of fact. "Just blood everywhere. I've opened that door there on the machine, and the blood just comes spilling out." She moves her arm to indicate a *Shining*-esque flood. "It's a lot of cleanup."

After pursing her lips and looking into the interior of the machine for a while, she decides it's good enough and hip-bumps the door closed.

The device is about the size and shape of a big trash can with wheels on the bottom. The top is sort of angled so it's looking up, and there are these big round knobs that look like eyes, surrounded by little twists and curlicues of tubing. There's a handle underneath shaped like a smile, and between the handle and the knobs, there's a little hole where the final length of tube disappears into the thing's insides. When she flips the switch, the machine whirs and clicks with satisfaction, and the blood spirals through all the tubes before getting sucked into that hole, which is obviously the feed hole of a deranged sentient bloodbot, and the knob-like eyes start to *spin*. It's like drinking blood makes it feel silly. When it stops sucking and switches gears, it makes a sound that's uncomfortably close to a sigh.

There's the sensation too, as the returned blood flows down into my hand and spreads through the network of capillaries there, of cold radiating through my fingers *from the inside out*, a deep chill that for most of human history was a real clue you were dying. I'm not saying the operation is unpleasant, just unusual. I'm not in any rush to finish. I wouldn't mind catching a little of *Interview with a Vampire*.

Let's be clear about one thing: it's funny to show a vampire flick in a blood center. But it also makes sick sense, especially if you're trying to figure out the price point for a pint, because nothing illustrates the preciousness of blood like the BII-type suspicion, held by people throughout history and across cultures, that something somewhere wants to suck it out of you.

The Babylonians placed plaques above their beds to ward away the donkey-bodied Lamashtu, and expectant mothers wore amulets to protect them against Lilith, Adam's first wife, the "flyer in a dark chamber," whose specific taste was for the blood of newborns. In Greece, they were wary of the bronze-footed Hecate, and on the banks of Kumpupirntily, Australia's blinding salt lake, they watched for the Ngayurnangalku, and near the Great Lakes of North America, Algonquian and colonist alike looked out for the windigo, his rags and buffalo horns hardly concealing his emaciated form.

And while it's easy now to brush off the concept that an otherworldly creature wants to steal your blood, it's hardly as if this concern was unwarranted in the past. There were blood markets of one sort or another even thousands of years ago, though they functioned differently than ours does today, existing mainly to feed the appetites of various deities. In Tenochtitlan, the gods Huitzilopochtli and Tlaloc could only be assuaged by spilling vast quantities of human blood down the white steps of the Templo Mayor, while Jehovah required the blood of Isaac as evidence of Abraham's fealty. During the Shang dynasty, the Shang king queried the needs of his god, Shang-di, with questions carved into cow bones.

辛酉其若亦汎伐

SHALL HUMAN BLOOD BE OFFERED ON THE DAY OF XINYOU?

He then prodded the bones with a bronze rod fired to such a degree that its heat caused the bones to crack, revealing Shang-di's will. There's no record of how the blood was extracted, but a grinning bronze axe was recovered from Tomb M1 at the Sufutun site, and we know the offerings were described as 汎, a "cascade." We know during the reign of Wu-ding, the god could only be placated by the death of nine thousand slaves.

Even more recently, back in the 1930s, in cities in Kenya and Tanganyika, a belief arose that white colonial firemen were stealing the blood of native Africans. Sometimes people were murdered on the spot, it was said, their blood drained into large buckets that the firemen ferried brimming back to the firehouse. Other times, they were merely abducted for later drainage. As one African politician said of a man who had been missing for decades: "We thought he had been slaughtered by the Nairobi Fire Brigade between 1930–1940 for his blood, which we believed was taken for use by the Medical Department for the treatment of Europeans with anaemic diseases." The belief was so common, in fact, that one of the terms for vampire in Swahili, *wazimamoto,* means "men who extinguish fire."

This vampire and human sacrifice stuff isn't a quaint aside to the quandary of what a pint of human blood is actually worth. As BII phobia and September 11 make clear, blood isn't just another resource. You can't just dig up more when you need it. The supply is constrained, not by a shortage of labor or materials but by the willingness of a human being to take an hour or two out of their day and have someone open up a hole in their arm. If you really want to get at an accurate price for a pint, you've got to get a handle on the way society has shaped people's feelings about being bled.

If the history of bloodletting is any indicator, a lot of people actually sort of like bleeding. Popularized by Hippocrates and Galen, the idea behind bloodletting was that by reducing the volume of blood in the body, the system was brought into balance, and various maladies—gout, fever, seizures, cholera, acne—were cured of their own accord. The procedure was a relatively simple one. A lancet or other sharp blade was used to make an incision through the skin and into a blood vessel. Occasionally, as a means of encouraging the reticent bleeder, a flame was burned inside of a glass that was then placed over the wound while still hot so that the air within the jar, creating a vacuum as it cooled, quite literally sucked the blood from the body. The placement of the bloodletter's incision was of great importance at times—the particular ailment of the patient, along with the position of the zodiac, dictated where the blood should be drained—though it was hardly an exact science, as evidenced by the fact that early bloodletting manuals are themselves often spattered with blood.

Perhaps the most famous recipient of this treatment was George Washington. He relied upon bloodletting throughout his life, even up to the morning of December 14, 1799, when he woke with a throat so sore he could hardly speak or swallow. Over the next ten hours, in addition to administering enemas of chlorine and mercury, his doctors drew off over three liters of blood from his circulatory system, a small fortune on today's market (though likely worth more now when one takes into account the £10,000 paid at auction for a few blades of grass stained by Gandhi's blood). He was a big man, probably a little over six feet, but in all likelihood, they drained off more than half the blood in his body, which reduction and ensuing malaise, as later scholars have pointed out, contributed to the consensus of those in attendance that the man said to be the father of the United States died in a state of peace and equanimity.

Less often recorded than the doctors' fatal prescription is Washington's own fervent belief in bloodletting. The former president, showing the plucky democratic resourcefulness of the nation he'd helped establish, had already performed venesection upon himself even before the arrival of medical professionals. It's a tricky operation for one person to accomplish on his own, and he'd enlisted the help of his overseer, even offering encouragement when the man showed some trepidation at slicing open the skin, saying reportedly, "Don't be afraid. The orifice is not large enough. More, more."

It's strange to think about the first president this way. It's hard to square the Crossing-the-Delaware idea of Washington—cape, saber, icy gaze—with the historical reality: a man earnestly and idiotically bleeding himself to death. It's not as though he was alone in this belief, either. Many in colonial America practiced bloodletting. John Adams, Thomas Jefferson, a who's who of founding fathers thought that sometimes a body just needed to be bled. And it's hard to come to terms with the fact that the truths we still hold to be self-evident—the equality of human beings and their right to self-government—were born in minds that believed it was a good idea to drain your own blood.

While Washington was very much of his time in terms of his allegiance to bloodletting, he was an outlier in the liberality with which he practiced it. Many doctors at the time believed that venesection, while effective, was too potent for certain patients: women, children, slaves, the aged, and the infirm. For all their occasional differences, it was with regard to bleeding the enslaved that Washington and Jefferson found themselves on irreconcilable terms. At Monticello, Jefferson's mantra was concise: "Never bleed a Negro." Washington, on the other hand, had been inspired to employ the practice so assiduously upon himself because he'd encountered such success bleeding his slaves.

I THINK REGULARLY OF AN ANECDOTE ABOUT A French doctor during the early days of transfusion. He arrives at a birthing room to discover a newborn girl lying still and lifeless, hardly different in color from the sheet in which she has been swaddled. The doctor sets to work quickly, incising first the arm of the father, then that of the child, then sewing the vein of one together with the other, so the father's blood runs into and through the body of the child, and the child's likewise into the father. Almost immediately, the girl's color returns. She opens her eyes, kicks and cries like any other newborn. What brings me back to it again and again isn't so much the miracle of the saved life. It's the physical connection, the way for a few moments their two hearts created a single circulatory system.

Though you're separated now from the recipient of your donation by time and distance and a great deal of flexible plastic tubing, this idea of a single shared circulatory system underpins every blood donation. It's what makes blood donations a radical gesture of equality, and it's what makes them, as well, a threat to ideas of racial purity or superiority. It's what makes donating blood in the United States of America such a strange experience.

For a long time, after all, the blood of Black Americans, in the eyes of the White population, had almost no value. Not long before Charles Drew organized Blood for Britain, the WPA conducted a series of interviews with former slaves, and there is hardly an entry in the archive where blood does not drip, pour, or run down the bodies of the enslaved. As Maggie Wesmoreland of Arkansas recalled of her former master, "He would strip me stark naked and tie my hands crossed and whoop me till the blood ooze out and drip on the ground when I walked." Or as Mingo White confided to an interviewer, the slaves in his part of the country were often whipped, as was he, for the simple offense of "prayin' for de Lawd to free dem lack he did the chillum of Is'ael." We know blood so covered ships in the middle passage that one witness, during his testimony before a Boston grand jury, described the scene beneath decks as "like a slaughterhouse." Despite antebellum doctors' insistence that slaves could not withstand bloodletting, their prescriptions for slaves too sick to work indicate otherwise: "nine drops essence of rawhide," one reads. Another recommends "oil of hickory."

And in December 1941, a few days after the attack on Pearl Harbor, Sylvia Tucker arrived for her first appointment at the newly opened blood center in Detroit and was informed, first by the desk nurse, then by a supervisor, that they could not take her blood. "Negro blood" was neither required nor accepted. The ad hoc policy had been quietly decided by the surgeons general for the army and navy earlier that year, in response to a letter from the chairman of the American Red Cross. "Your query regarding the use of negro blood as far as the Navy is concerned, is a very simple one to answer," wrote the surgeon general for the Navy. There was no use for it. As a young man named Robert Carlyle Byrd wrote a few months later, "Would a white soldier, if wounded on the field of battle accept blood possibly transfused from Negro veins? As for myself I expect to fight in this war and I would rather receive blood from a monkey or a guinea pig or suffer death if offered no other choice."

By early 1942, however, the informal policy of excluding Black blood from the national blood supply had already become untenable. Girl Scouts in Brooklyn sent letters of protest to their local Red Cross blood center. Editorial and political cartoons appeared in the *Philadelphia Inquirer* and the *Baltimore Afro-American*. The chairman of the Red Cross, Norman H. Stimson, was flummoxed by the issue. A former diplomat, he had worked to restructure financial markets and establish reasonable reparations following World War I, negotiating an agreement among the nearly thirty countries involved in the Paris Peace Conference, but as he told a friend that year of the issue, "I have never dealt with a problem more loaded with dynamite." In January 1942, after Stimson organized a series of talks with many of the principals involved, the Red Cross agreed to a new policy. The blood of Black donors would be accepted, but it would be labeled AA, Afro-American. This way it could be segregated from the general blood supply.

The decision was not well received. Civil rights leaders met with officials from the Red Cross. The Communist Party of Cuyahoga County, Ohio, rallied 3,500 people in protest and sent a letter to the mayor of Cleveland. And a pair of high school students ambushed the secretary of war in his office. "They pleaded against the discrimination," he wrote in his diary, "which they had heard the War Department was making against negro blood." In an act of sublime and defiant patriotism reported by Langston Hughes, many Black women, "light, and to the eye, quite near-white Negroes," protested this policy by "putting on their light makeup and bearing their lovely arms to the needle."

In April 1942, after only three months with the Red Cross, Drew resigned from his post, returning to Howard University to teach. It is hard to imagine how difficult it must have been for him to watch the national blood banking system, in no small way his creation, twisted and deformed by prejudice. His anger and his confusion are palpable in the historical record.

"It makes no sense," he told the American Serological Association of the decision to segregate the blood of Black donors. "First, it has no basis in science, second, it is humiliating, and third, we need the blood."

In 1944, when the Red Cross confidentially reviewed its data on Black donations, they discovered Drew was not alone in his feeling of humiliation. At the time, though Black Americans were only 10 percent of the population, they were avid participants in the national war effort. They bought defense bonds, they planted victory gardens, and of course they fought and died in the war. They should have accounted for a substantial portion of donations, but the Red Cross found that Black donors had been responsible for less than one percent of blood collected.

Though the Red Cross officially stopped segregating blood in 1950, that disparity in participation remains to this day. Black donors are still underrepresented in the donor pool. When it comes to blood, the living heritage of slavery isn't just a topic for wan reflection at private liberal arts colleges in the Northeast. It's a daily crisis. If you're interested in the Slonim-Wang-Garbarinonian ideal of efficient national blood collection and distribution, one of the most galling inefficiencies has nothing to do with flying Duffy-negative people across the world to donate a pint. It's the chronic shortage of a specific subtype of the Rh antigen system, a kind of blood known as R_0. While there's always a lot of demand for the most common blood types like

O+, and while a rare blood like Rh_{null} is invaluable in specific situations, R_0 has the misfortune of being both rare and in high demand, and the best source of R_0 is the Black population. As one doctor said with urgency, "We need Black blood."

AT MY BLOOD CENTER IN SANTA FE, MY DONATION IS almost done. I eat some cookies. What I love most about the blood center isn't so much any sense of pride or decency as the steady clinical hum of an operation proceeding in its precise Drew-ian order. There can be a wildfire beyond the ridge, a failed coup simmering in a distant capital, but in the blood center, all is calm. Blood does not reach Allah, as the Quran says, only piety. I have worked on dairy farms connecting and disconnecting endless sucker cups to endless udders, and the demeanor of the blood donor in situ is not dissimilar from that of the cow being milked. The thing will be done when it's done. You know the bag of blood filling at your side will be of use to another's body, will clot if they cut their finger, will carry breath to their heart, their liver, their brain, will run quick in their veins if they see a lover or a friend, and will bring to their lips some soft but essential hello. But you don't think about it too much. It almost feels like it isn't about blood at all.

IV

Semen

> **Terms**
>
> n. seed, spunk, cum, spooge, wad, jizz, juice, skeet, squirt, nut, load

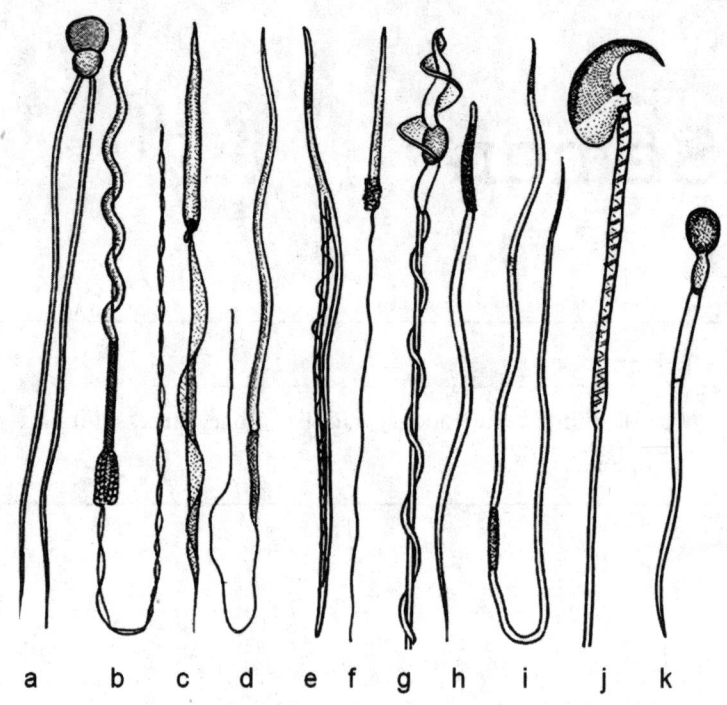

Variations in sperm structure across different vertebrates: (a) Toadfish, (b) Elasmobranch, (c) Toad, (d) Frog, (e) Salamander, (f) Lizard, (g) Fringilla, (h) Domestic fowl, (i) Monotreme, (j) Mouse, (k) Man; *Chordate Zoology*, P.S. Verma, 1965

Biological Prologue: Sperm Journey

The first thing to know is that, while we often think of ejaculation as a sudden act, the male orgasm is only a small part of a much longer and convoluted process. Every sperm takes about seventy-four days to mature, and because that whole operation is jammed into the volume of a testicle, the architecture of the testes, especially in humans, is mind-bogglingly labyrinthine.

Production starts in the seminiferous tubules, essentially a thousand long, squiggly U-shaped tubes radiating out from and returning to a point at the rear of each testicle. Their interior surface is covered with conical structures resembling little Christmas trees, the Sertoli cells. The Sertolis' fat ends are planted in the wall of each tubule, while the narrow tips extend into the tube's open center. What lends the Sertolis the appearance of Christmas trees is that they're decorated all over with sperms in various stages of development, which they're nursing into being. At the bottom, closest to the wall of the tube, are undifferentiated spermatogonia stem cells. Though we don't know what controls the timescale, in regular intervals, a pulse of retinoic acid washes down the tubules, causing these proto-sperm to mitotically divide and move one layer up the tree. Another pulse, another mitotic division, and they move up another layer. A few more pulses, and as they reach the top of the tree, their tails begin to grow, stretching out and waving in the lumen-filled middle of the tube until their tails are ten times the length of their bodies. At last, the Sertoli cytoplasm recedes from around the head of the flagellate, microtubular rings squeeze the sperm like a corset, and off it pops.

Following disengagement from the Sertoli, the sperm don't swim. They're washed away by the current in the tubule. Down they go to the branching anastomosing rete testis, a sort of river delta of crisscrossing channels where the many sperm from the many tubules mix, and from here, they pass through the efferent ducts and into the epididymis along with millions of their companions. Due to the sheer quantity of sperm and the need of many, if not all, men to calm their niggling fears of inadequacy with notions of insuperable virility, there is a temptation here to think of these massing sperm as a sort of expeditionary force, rucksacks packed and ready for an invasion, but this is all part of the gung-ho macho ethos of impregnation. Rather, they are what they are: a bunch of largely insensate microscopic flagellates, going with the flow.

They are swept forward into the cilia of the epididymis by the absorption of fluid, down, down, down, along the backside of the testicle. This epididymal voyage alone, a traverse the length of one testicle, takes from two to six days. It seems a surprisingly long trip considering that the testes, usually measured using a rosary-like string of beads known as a Prader's orchidometer, are rarely much longer than two inches. The duration of the trip isn't too surprising, though, when you consider that the epididymis of the human male is so delicate and twisted up that its exact length has not yet been determined. Estimates place it in the range of eighteen to twenty feet, a length which, if true, would put the small intestine to shame.

Finally, the sperm arrive at the cauda, a bend at the bottom of the testicle that acts as a storage site. Still, they don't move. It is almost as if they're sleeping. The passage through the epididymis has concentrated them in what is called "sperm packing." Here, bathed in proteins, electrolytes, and small organic molecules, they are ready to depart the organism. Should ejaculation occur, muscles in the groin will contract, and these sperm will be expelled from the cauda and flushed down the vas deferens. They'll join with the seminal vesicle's fructose, phosphorylcholine ergothioneine, ascorbic acid, flavins, prostaglandins, and bicarbonate; the prostate's spermine, citric acid, cholesterol, phospholipids, fibrinolysin, fibrinogenase, zinc acid, phosphatase, and prostate specific; and from the bulbourethral gland, the gland of Littre, and the lacunae of Morgagni a clear mucoid fluid. And the whole mixture will be sent down the urethra and emerge from the penis as an off-whitish fluid called semen: specific gravity, 1.028, barely denser than water; pH 7.4, slightly alkaline; average volume, three milliliters; a mass of a few paper clips.

While anything in the range of white to gray is considered normal, semen may take on many hues. Yellow can signal contamination with urine, though it may also result from a carrot-heavy diet. Clear semen indicates a low sperm content, and reddish-brown or darkly flecked semen suggests the presence of blood, a condition known as hematospermia, possibly the result of a tumor but more often of a more benign derivation. Trauma to the groin is a frequent cause. Texturally speaking, semen is unique. Though emitted as a liquid, contributions from the seminal vesicle cause this fluid to coagulate immediately after ejaculation into an irregular gelatinous material with one hundred times the viscosity of its most liquid state. Over several minutes, it then reliquifies due to the activity of proteases from the prostate, allowing the sperm to exercise their motility.

The quality and quantity of the sperm within this semen are affected by everything from diet to genetics to age, obesity, environmental pollution, and hot tub use. Though the mechanisms involved have not been established, studies suggest olive oil, fiber, and carrots increase sperm quality. Smoking, on the other hand, reduces semen volume, sperm counts, and motility percentage. It's worth noting, though, that very few sperm are ever truly functional. They have heads that are too large or too small or strangely shaped, or their tails are crooked or coiled or come out sideways. In fact, only about 10 percent qualify as normal healthy sperm. A sperm analysis—which assesses sperm number, density, vitality, and motility—is the best way to assess the health of one's sperm, but some sense of fertility can be derived by a simpler measure: the volume of the testicles. The testes are not static organs. At puberty, they grow, peaking in size around the age of thirty and remaining there until sixty, at which point they begin to shrink. They also change in size in response to the hormones circulating in the body. They are, in andrology-speak, clinical markers of our hormonal and spermatogenic functions. And while larger testicles have absolutely nothing to do with semen volume (which is largely determined by the fluid from the seminal vesicles), they do correlate with healthy sperm creation.

Sperm, it should be mentioned, are by no means uniform across the animal kingdom. The heads of mouse sperm are shaped like executioners' axes, while the sperm of the genus *Fringilla*, a group of Old World finches, feature a helical spiral along their entire length similar to the ribbonlike tentacles of sea jellies. Perhaps most remarkable, the sperm of *Drosophila melanogaster*, the common fruit fly, have a tail almost forty times as long as that of the sperm of humans. (If the gametes of men were to have proportionally long tails, their sperm would be about as long as their arms.) Sperm do not move in the same manner either. With its immense tail, *D. melanogaster* often locomotes tail-first, dragging its head along behind. The male bedbug, on the other hand, *Cimex lactularius,* pierces the abdomen of the female with a stylus, and his sperm move through her circulatory system. Even the sperm of humans do not move as is commonly assumed. They don't propel themselves forward like a fish with a side-to-side movement of their tails. Instead, they move their tails asymmetrically, resulting in a corkscrew motion, similar to the spiraling movement of a frolicsome otter.

When we think of these sperm being jettisoned from the comfort of the cauda, rushed down the urethra by the internal musculature of the body, and sent out into the exterior world with its variations in temperature and acidity, they *do* seem a little like soldiers being disgorged from a ship onto some foreign beachhead, but if we must anthropomorphize them, I prefer to think of them as newly matriculated college students, setting off with little more than a spiraling form of locomotion and a few electrolytes in search of they know not what.

EARTHLY MATERIALS

Excessive Artificial Stimulation

I am going to be GREAT.

u/Melodic-Tax-3821

181 days

To ejaculate, in the eyes of the Judeo-Christian god, is a sacred act. One's seed shouldn't be scattered vainly, the early Christian theologian Clement of Alexandria writes. But there are limits to its sanctity. Semen is not the body of Christ, insists Epiphanius, discussing his attempted recruitment in the fourth century by the Borborites, a sect of gnostic Christians who reputedly took the substance as their sacrament: we should not eat it. The much-discussed story of Onan, who was killed by the Lord in the Book of Genesis, doesn't tell us much. Along with being used to make the case against masturbation, his crime—the spilling of his semen on the ground—has been interpreted as a parable against contraception, against oral sex, against anal sex, against the withdrawal method, against lust, against greed, and, according to scholars of the original text, as a defense of the institution of levirate marriage, wherein a man is required to wed and impregnate his brother's widow.

Though many turn to nature for guidance or reassurance, the animal kingdom doesn't have much to say about the ethics of ejaculation. Orcas make amorous advances on their human trainers. Lions mount their feed containers. In the Antarctic, certain penguins ejaculate on rocks that look like penguins, and otters in California sometimes copulate with seal pups, carrying on even after the pups' decease. With a faraway look, Labrador retrievers thrust themselves upon the shins of dinner guests. And while some use the animal kingdom to make the case that fertilization is the one, the only, and the ultimate goal of ejaculation, it's instructive to remember the unassuming female Muscovy duck, whose mazelike vagina, with its manifold cul-de-sacs, evolved specifically to stymie and divert that goal, in the process driving the tortuous penile evolution of the male Muscovy.

In popular culture, the disposition of semen follows no well-defined schema. Instead, as in William S. Burroughs's *Naked Lunch*, it seems to drift in ribbons through the collective psyche. There's the enormous quantity of pornography on this planet, of course. Male ejaculation is the raison d'être of much of it, sending semen flying willy-nilly across screens worldwide. In Noah Baumbach's mid-2000s coming-of-age film *The Squid and the Whale*, the character of Frank wipes his semen on the spines of the books in his school library, while James Joyce, the most literary of writers, is more interested in ejaculating on the face of his wife, Nora Barnacle, writing to her in December 1909, "I feel mad to do it in some filthy way, to feel your hot lecherous lips sucking away at me, to fuck between your two rose-tipped bubbies, to come on your face and squirt it over your hot cheeks and eyes." Life begins sometimes to feel like one of those pornographic videos on the internet—ejaculation forever looms—and it reminds me of something I once heard a mom say about living with her three teen sons. The central psychological premise of raising a boy, she said, is the recognition that everything has at least some semen on it.

That seems pretty accurate, based on the accounts given by my friends in middle and high school. My friend TJ masturbated all the time and everywhere, though he preferred the bathroom at church, thus escaping both his sexual appetites and the sermon preached against them. Stubbs, an atheist and functionalist, preferred convenience above all else. He masturbated only at home, directly into the toilet, which seemed, he sometimes mused, specifically designed to carry off the product of his vigorous self-depredations. My friend Kahlil used to take his mom's hot water bottle, lather it in jasmine lotion, fold it between two cushions of the couch, and have his way with the afternoon, while one of my band friends, Ricky, most enjoyed the ritual of the act, following the same daily routine without deviation during the hour and a half between the end of our school day and the end of his mother's workday. He'd lock the door behind him when he got home, drop his bookbag on the couch, grab a Capri Sun out of the fridge, and microwave a Hot Pocket, which he'd eat while his mother's computer booted up. Then he'd angle the venetian blinds, open up a search engine—AltaVista, he said, kissing his fingertips—and type "boobs." He kept up the routine for years, through the end of middle school and the beginning of high school, through two girlfriends, several dances, his driver's license exam. He'd probably still be doing it today except one afternoon, blowing on his Hot Pocket, he entered the computer room and found his stepfather, home early from his job at RadioShack, leaning over the keyboard with his pants around his ankles.

We were sitting in the band's practice room when he told

me this. I was leafing through our sheet music. Ricky pressed the spit valve on his sousaphone and blew into the mouthpiece. A few dribbles of saliva fell to the carpet, and he shook his head. "He never even cleared the cache."

When we got older and Kahlil's girlfriend moved in with him, he encountered a novel stressor. It didn't have to do with who would wash the dishes or what TV shows they'd watch or how to make emotional space in his life for this other human being. It was that he no longer knew where to masturbate. It was still the same apartment he'd been living in for a year—the same couch, the same bathroom, the same desk, the same computer—it just suddenly felt *weird*.

Part of the reason is that masturbation's gotten a bad rap historically. Aristotle believed it was foolhardy to emit the precious fluid except when absolutely necessary, as it would drain one of vital energy. And the Calvinist Swiss physician Samuel Auguste Tissot argued in *L'Onanisme* in 1760 that masturbation would lead to "post-masturbatory disease," a condition elaborated by later writers to include infection, sexual dysfunction, insanity, neurosis, neurasthenia, epilepsy, harm to the genitalia, fetishism, homosexuality—in other words, "complete degeneration of the masturbator." Voltaire and Rousseau agreed: most human diseases were attributable to masturbation. Even Newton, fearful of eroding his mathematical powers, subscribed to the idea that masturbation would affect his mental processes, in which frame of mind the universe as we popularly understand it—a system of attractive bodies bound by angle and momentum—was conceived.

If you were never a thirteen-year-old boy, or even if it's been a while since you were, the history of the ethics of ejaculation might seem pretty far down the list of issues worth caring about. But despite several millennia of intense discussion about where and when it's cool to ejaculate, despite the best efforts of some of the world's most active and deliberative minds, the matter still isn't settled. At this very moment, in a not-so-distant corner of the internet, on a sub-Reddit forum known as r/NoFap, the debate is still happening. Twenty-four hours a day, seven days a week, hundreds of thousands of young men, and one or two women, are eagerly discussing how to stop masturbating, and it might be that for the first time in history, the discussion is actually important.

> Had this thought during post nut clarity, I think it makes sense to share it.
>
> u/Tuber993

FOR THOSE WITHOUT MUCH EXPERIENCE OF THE WEBsite Reddit or the redditors who use it or the Reddiquette they follow, it's just a giant virtual bulletin board where people post thoughts and images and comment upon them. There's nothing super innovative about the concept except that the bulletin board is subdivided into a collection of smaller bulletin boards. They're called subreddits or subs. There's r/funny, a discussion forum for funny things; r/todayilearned, a forum where people relate facts and lessons recently discovered; r/DIY, a forum for do-it-yourselfers. And then there's r/NoFap, a forum for people trying to stop masturbating to internet porn.

Here's how things work on a subreddit: making use of a great deal of Reddit-specific shorthand, a redditor (the OP—original poster) initiates a discussion (a thread) by posting their thoughts (IMO—In my opinion or IMHO—In my honest opinion) or asking questions (ELI5—Explain like I'm 5) or by simply opening themselves up for interrogation (AMA—Ask me anything or AMAA—Ask me almost anything). This post is read by other redditors, a decent portion of whom are deeply insulated from the nondigital world (the average redditor is what many on Reddit might call a neckbeard: young, male, mobile-savvy, and with some college under his belt, though many are young enough to still be in high school), and these redditors respond with their own comments and also by upvoting or downvoting posts and comments, thus increasing or decreasing the visibility of certain threads and certain discussions within threads and resulting in points for the redditor who wrote the post or comment. The result? An often contentious but still self-reinforcing dialogue that sometimes results in an echo chamber–like atmosphere of shared values (what is sometimes known as a circlejerk or hivemind), perhaps the most remarkable example of which is the very terminology I'm describing, the esoteric involute lingo fifty million or so redditors use to communicate daily.

The subreddit r/NoFap occupies a special place in this world in the same way the last stall with the broken door occupies a special place in the restroom at the public pool. It started about ten years ago as an online space where men, usually youngish men, could digitally congregate and encourage one another in their efforts to stop masturbating to internet porn, but it's developed into something else as time has passed. It's now become a clearinghouse for all the questions and confusions a person might have about their organism's desire to self-gratify, with a community of like-minded folks ready and willing to offer encouragement, perspectives, wisdom, and counterfactual medical advice.

If you have a recurring dream that you're typing "porn" into Google, you ask the r/NoFap subreddit what to do. If your wife told you she's ready to have children and now you can't seem to get an erection no matter what you do, you confide in the r/NoFap sub. If you think the woman who orders her taco bowl from you every day is flirting with you, or the jogger seen from afar in the woods, or your stepsister, or your therapist, or Siri, you take these suspicions to the r/NoFap sub. If you're making tea and your penis rubs against the cool bull-nosed edge of the kitchen counter, if you just masturbated to a pile of used clothing you bought for that purpose on Craigslist, if you've been wandering lost in a "demonic realm of fap," if you can't sleep on your stomach, if your penis seems to be shrinking, if you're currently wearing a sports cup around the house, chances are you're on r/NoFap. Though women occasionally join in the conversation, in many ways it's a distillation of the world Reddit already is: younger, maler, more digital, more lonely. It's the Reddit of Reddit.

r/NoFap comes replete with its own subreddit-specific jargon, which, while a little onerous to learn, does provide a fascinating window into the minds of those on the sub and the concerns that occupy those minds.

fap: to masturbate, etymology unclear, supposedly onomatopoeic

NoFap: technically, a registered trademark owned by NoFap, a Pittsburgh-based company helmed by Alexander Rhodes, the man who accidentally created the NoFap subreddit forum by writing a Reddit post about his own struggles with internet pornography; according to Rhodes's organization, the term specifically describes "a comprehensive community-based porn recovery website," but for all practical purposes on the sub, it means not fapping, especially to internet pornography

nofapper: a member of the r/NoFap community

PMO: porn, masturbation, and orgasm, the trifecta of no-no's to many on the sub

PIED: porn-induced erectile dysfunction, a symptom, according to some, of too much PMO

reboot: abstaining from certain sexual behaviors (e.g., PMO) for a set time in the hopes of restoring oneself to a prelapsarian state, occasionally done as part of a collective challenge, as in #NoNutNovember

streak: number of days without PMO

#NoNutNovember: why many Thanksgivings have gotten more tense

hard mode: reboot without any sex

monk mode: no precise definition exists for this "hard mode of hard mode," but it possibly entails no interaction with women, no internet (except r/NoFap), no sugar, lots of weight lifting, and spiritual centering practices; socializing is, according to some, "acceptable"

DeltaFosB, or ΔFosB: a lingering protein implicated in addictive behaviors, believed by some fapstronauts to bedevil their reboot, hard mode, and monk mode efforts

peeking: looking at porn without fapping

edging: fapping without orgasming

relapse: reboot failure (i.e., the O in PMO)

While that glossary covers the basic lexicon, there's a lot more, and more revealing even than the sub's language are the everyday topics and terms that seem to come up again and again. r/NoFap posts often include a tag (motivation, advice, victory, slipup prevention—urgent!, relapse report), and there's a lot of talk of lust, shame, wickedness, demons; lots of *The Lord of the Rings* memes; lots of extensive references to dopamine, brain fog, penis size and venation, prolactin, stress, anxiety, depression, isolation; and lots of doctored stills from *The Office*. Also, many r/NoFap redditors attach a counter to their username—167 days, 12 days, 8 days, 200 days, 2 days, 0 days, 0 days, 0 days—to indicate how long it's been since they last fapped.

You might have the sense that this is all a little much, that this is a part of the world you don't need to understand, in fact one that you'd prefer not to know exists. You're not wrong. The NoFap LLC itself has rightfully expended a great deal of effort distancing itself from the more extreme elements of the sub. It's a scary part of the internet, this world where young men talk online about porn in an effort to think less about porn and be less online. There are some deranged concepts of masculinity. There are anti-Semitic porn conspiracies. There are truly frightening notions of the roles women should play in society, backed up by ramshackle conceptions of biology and history. Most profoundly, there are just a lot of young men struggling to understand their sexuality in a free-for-all forum of thoughtful and supportive nincompoops. (One of the recurrent concerns on r/NoFap, for instance, is whether wet dreams constitute a relapse.)

The urge to flee is not unwarranted. There's a lot of world out there. You could be watching rocket launches at Cape Canaveral, painting watercolors of the Washington Monument, strolling amongst thickly trunked sequoias, researching the migratory patterns of sperm whales, laughing along to season four of *The Great British Bake Off* where Christine tries her hand at spotted dick pudding, watching a volcano explode, eating a pickle. You could have a conversation with a woman, ask her how her day was.

But the internet doesn't stop existing just because you look the other way, and while these discussions happen largely online—just pixels, as they say on the sub—they continuously intersect with the world in ways both small and large. Sometimes, those intersections seem positive: the man who decides to reevaluate his habits and pursue a different kind of relationship with his fellow human beings. Some intersections are harder to evaluate, like the man who's convinced not masturbating is giving him muscles and making women smile at him or the man who's channeling his urges into building a sneaker collection. And sometimes—the man who brings a gun to the yoga studio, the man who brings a gun to the massage parlor, the man on Day 48 of his reboot who writes, "The urge has changed to something sinister"—sometimes the online world of aroused and frightened young men intersects with this one in ways that are very real and very terrifying. There are nearly nine hundred thousand Redditors on r/NoFap out there. It pays to have some sense of what's going on.

There's another important layer to it, though, too. If you interact with the internet in any way, as you almost certainly do, you're already deeply tangled up in what's occurring on the r/NoFap subreddit.

While porn isn't the same as the internet, it did play a large part in building it, enough that you can't accurately relate the history of one without describing the history of the other. You might believe the internet is mostly a fun place where you look at photos of your friends and family and watch videos of baby elephant birthday parties and shop for mid-century modern shower curtain rings, but think about the underlying structure of the internet as you now know it—communities discussing things in a semi-anonymous bulletin board fashion complete with pictures and video and banner ads that, when clicked, lead you to megastores composed of affiliated sellers where your credit card or your PayPal account is waiting and ready to be charged. That's the internet porn hath wrought.

Almost the entire architecture of the space was in some way influenced by the drive to masturbate. Event Horizons BBS, which was an early Bulletin Board System, was a clearinghouse for pornography, and by grossing over $3 million annually in the early 1990s, it offered some of the first hints of the many ways money could be made online. With its anonymous handles, chat function, and image sharing, BBS also influenced the design of AOL and much of the structure of social media that followed. One man, Ron Levi of the website Cybererotica, helped pioneer everything from pay-per-click and pay-per-acquisition banner ads (which, with Google's addition of tracking and targeting, have become the mother lode of internet profitability) to double opt-in email subscriptions (still a central fraud deterrent) to affiliate sales networks (now the Amazon model). Meanwhile, some of the first digital currency, "e-cash" as it was then called, was introduced by a company called CyberCash, with its CEO noting that the investment of "adult entertainment will be substantial in the early stages." CyberCash, through a series of post–dot com bubble acquisitions, is now part of PayPal.

The way the internet indulges our sense of anonymity, the way it caters to our impulsivity (*just one more episode and then I've got to get to bed*), the way it satisfies both our voyeurism (FAIL compilations) and our exhibitionism (Insta), the way it drives us to brainstemmy, rapid-fire styles of communication, and the way it monetizes all of this through clicks and subscriptions and views and conversions—whatever pathway of neural rewards drives a human being to ogle a screen, the internet has followed it.

Xvideos.com and Pornhub.com, the leading pornography sites worldwide, rival the traffic of X (formerly Twitter) and

Instagram, and if the portion of visits made using a mobile phone is a marker of a website's positioning for future growth, these two sites—both of which have about 90 percent engagement via mobile—dwarf larger entities like Facebook and Amazon. Pornhub, as of its last accounting, hosted well over a million hours, literally several lifetimes, worth of pornographic videos. It has over forty billion visitors each year, tens of thousands of visitors each minute, tuning in from across the United States and the world. It is an undeniable bulwark of the internet—in other words, one of the main buttons we push to stave off the ever-looming threat of ennui. When Meta's suite of sites went down a little before noon on October 4, 2021, Pornhub's traffic immediately jumped. It doesn't approach the behemoth that is Google in terms of monthly page visits via mobile phone in the United States, but Google is a gateway, not a destination. With close to a billion monthly visits, Pornhub comes in at a respectable number seven, just behind Reddit.

People come to the r/NoFap subreddit because they feel they are being bodily devoured by this internet, and so the sub has ended up being sort of the internet's contradictory conscience. It's the place where young men attempt to separate themselves from the digital world by engaging with a poorly informed digital community in the hopes of getting upvoted. It's where the internet—your internet—eats its tail.

And look, a lot of it is dumb, like real dumb:

> For some reason when I started fapping, it felt as if the joy in me had fled like a frightened child who saw a

ghost. My mood fell from the mountaintops down into the deepest darkest valleys.

u/TheSleepingDonkey

Your sperm could create life and if you hold it, it brings life within you. It is your seed and with that seed, you could grow a tree that even the most beautiful birds could rest on.

u/Melodic-Tax-3821

Mid-30s male. Cold genitals no longer an issue.

u/domilalk

But every so often someone asks a question, as the redditor u/abbayyoo did in the run-up to #nonutnovember, a question that doesn't seem on its surface either insightful or interesting, that in fact seems utterly idiotic, but that question, for all its grammatical deficiencies and moral ellipses, somehow accidentally strikes at the very heart of this project to reclaim our existence from the internet.

Just had sex with a hooker

Hello guys , I'm on no fap journey currently two weeks. So I had an urge since yesterday but went and relieved myself with a hooker today. So my question is, does having sex with a woman, hooker or not help reverse the effects of masturbation? Waiting for your reply thanks

u/abbayyoo

In some ways, what happened after u/abbayyoo's post was textbook Reddit: a rapidly evolving, wholly self-contained world of largely unnecessary commentary. Numerically, it was a far cry from the sort of traction posts can get on other subreddits, and even on r/NoFap it didn't get the same level of attention as, say, a video of a heavily muscled shirtless guy doing dumbbell presses in what appeared to be a gym at an airport hotel while giving a motivational carpe diem–type speech into his phone. But something began to happen after u/abbayyoo posted this question, some magical combination of buttons was pressed in the ensuing back and forth, and it became a discussion not just about pornography or masturbation but also about something more metaphysical in nature. As u/The_New_Renegade would eventually note of the conversation, in what felt very much like a farewell to the entire r/NoFap sub, it was "like civil war or something."

> Truth be told, I'm scared out of my mind. Penis and balls are fine.
>
> u/Snowrend

WHILE YOU MIGHT EXPECT REACTIONS TO BE NEGA-tive when someone brings their story of paying for sex to a forum of people struggling to control their own sexual urges, one of the uncanny things about the r/NoFap subreddit is that, for all its weirdnesses, the place is actually pretty community-minded. While folks have strong opinions about porn, masturbation, sex, power, god, love, and DeltaFosB, many comments tend to be genuine, straightforward responses where users emphasize they are only providing their own perspective. The first response offered a great example of this:

> I doubt it reverses it. At least it was real sex, though going the hooker route wouldnt be how I would do it.
>
> u/Tiki-Light (16 days)

It was a solid reply, nonjudgmental and to the point. But it left unanswered the more difficult and nebulous question of whether paying for sex was acceptable behavior for a person trying not to masturbate, and other subredditors were quick to offer their thoughts, with u/carsww and u/soyboy2000 establishing opposing camps:

> I think you're good. The point of nofap is to stop having sex with your hand.
>
> u/soyboy2000

> Isnt this just masturbation with extra steps?
>
> u/carsww

At which point, everything blew up.

It might not be immediately obvious if you haven't spent more time than is really healthy researching fapping and lurking in the corners of the sub to see what exactly gets talked about, but this wasn't standard fare. This wasn't a conversation about how best to stop watching porn. It wasn't practical tips for avoiding arousal at school or work. It wasn't somebody asking for distractions as they struggled to just not type "cumpilation" into a private tab on their mom's phone. This was fundamentally an existential debate about *the nature of the subreddit itself.* Was r/NoFap about not masturbating to internet porn? Or was it something less easily defined? If sex with a sex worker wasn't sex, then what *was*? Was there such a thing as what David Foster Wallace once called "the felt reality of love," and what did that mean and how did one go about it? You can imagine how this sex-can-actually-be-a-form-of-masturbation notion might drive a wedge into an already divided masturbation sub.

A substantial back and forth between the two sides followed. There was debate on the difference between porn addiction and sex addiction. There was discussion of the role of fantasy in both watching porn and paying for sex. There was a lot of talk about the practical logistics of prostitution, which many seemed startlingly up to speed on. There was even this small but delightful aside with the OP:

how much it cost you btw?

u/Last-Donut

It's actually cheap

u/abbayyoo

> Dude don't come advertising prostitution in a porn addiction forum.
>
> u/Drizzle_D

But then u/King_Mandias (3 days) started probing into the exact distinctions that might make paying for sex just another kind of masturbation while other forms of sex were OK.

> So tell me how having sex with a hooker is any different from having sex with a casual one night stand you met on dating apps or something other than paying money? Or do you believe you should only have sex with your romantic partner? Genuine question.

The responses ran the gamut. There were those, like u/libbyson (160 days), who didn't believe there was any difference between prostitution and casual sex, but there was also u/ColdLake95 (677 days), who noted, with the troubling, sweet naivete that only comes from nearly two years of not masturbating:

> I had sex with a few hookers. There's a tiny chance you can get mutual pleasure where she also cums with you. You have some sort of small connection but it's something.

There was some dopamine talk, a short rant on prostitution and capitalism, but the consensus at the end was that there *could* be a difference between having a one-night stand and sleeping with a sex worker. It had to do with your own approach to the act of sex. Was it something you did *with* another person or *to* them? Did you care about the other person's enjoyment as much as you did your own? If such a thing were possible, did you care about their enjoyment *more* than your own? It didn't automatically become sex, by this reasoning, just because another person was there. There had to be an air of mutual concern. This obviously bypassed a lot of fun and worthwhile ideas of submission and dominance, as well as ignored some very important legal and technical definitions of intercourse, but no matter which side of the schism you were on, these were reasonable distinctions being made about the kinds of intimacy possible with other humans.

But then the self-control faction chimed in. As you can imagine, self-control is a topic near and dear to many on the sub. Like the rest of us, they've got phones in their pockets, and those phones provide an instantaneous portal to a world of porn larger than any human being could ever hope to circumnavigate, and occasionally—anybody can understand this, whether they watch porn or not—occasionally your phone is suddenly in your hand and you don't know how it got there. Something got the phone out and opened the browser, but it doesn't feel like it was you.

u/sbt178 (0 days) was ready to argue for self-control as a central pillar of not masturbating, posting a comment, which I will only gloss over here because the reproduction of such an earnest word salad brings me physical pain. What u/sbt178 did was basically imagine a conversation with his own brain, in which he told his brain "sex need hard work to 'earn.'" Orgasm, he explained, was a reward for the brain, and both porn and paid sex offered that reward too easily. You needed to "'earn'" sex by transmuting that sexual energy into productive habits—i.e., working to make yourself datable and building the "emotional empathy required to date someone." "NoFap," he concluded, just sounded catchier than "'No Excessive Artificial Stimulation.'"

With this comment, to use D. H. Lawrence's favorite euphemism for ejaculation, the thread had reached its "crisis." People loved the concept of excessive artificial stimulation, and with a great surge of upvotes, u/sbt178's comment shot to the surface of the whole conversation. And I don't know. In some ways, when you give it even the most cursory consideration, it's deeply, deeply, deeply screwed up. Even ignoring the fact that u/sbt178 spoke to his brain the way Tarzan might communicate if Tarzan used air quotes, the idea of anything as a reward for empathy seems to necessarily preclude empathy. But on the other hand, giving u/sbt178 the most charitable reading possible, there's a chance he was on to something. Human relationships rely on mutual trust, and building that trust requires a huge amount of work. It's daily and it's unending and it's important, and nobody ever teaches you how to do it, you're just supposed to pick it up on the fly. No, ejaculating inside another person isn't the reward for that work, but for people who care about other people, sex can be one of the many culminating acts of trust in this life. It's a reward in the same way seeing O'Keefe's Abiquiú studio is a reward for a person who's studied painting. It's not a reward for the work, it becomes a reward *because* of the work.

As the sufferers of postcoital tristesse sink into hopelessness, the conversation devolved at this point into some very PG name-calling ("cretin," "dumbass") and abortive side chatter ("People like u are ruining nofap"), but not before u/NatsuNoMercy (29 days) submitted one final notion of sex: it's "sort of like building a physical relationship with the other person as you explore their body to find new ways to please them."

> You need to have a greater good, something that you can sacrifice and orgasm for.
>
> u/TetsuJake

IT WAS A LOT. AND IT WASN'T WHAT I'D COME TO EXpect of a forum devoted to masturbating less, but in its own way, it was thrilling. Yes, there was little familiarity in this crowd with biology, anthropology, psychology, or women. Yes, they couldn't seem to grasp that *hooker* is a pejorative or that describing sex with the phrase "relieved myself" may merit therapy. Yes, these were a bunch of dudes who just wanted to have sex with another person real bad, but involved in that want were questions about when enjoyment becomes addiction, how one creates meaningful relationships, and what is real and what mere pixels—questions of need, control, love, and reality that are pretty fundamental to *any* internet-connected person's life. They were talking about the challenges of defining and creating meaningful human relationships in a world where physical interactions are rapidly being subsumed by digital ones. Replace ejaculation with whatever internet tendency most bedevils you, and you've got the groundwork for some serious personal reflection.

Did you spend two hours last week researching the best drill or the best pillowcases or the best toilet bowl brush? Did you spend an entire car ride liking Instagram or X posts of people you've never met or haven't seen in years while your spouse drove on in silence? Have you been playing a great deal of mahjong online? Did you spend the better part of a year cackling in the shadows of Reddit's most disturbing board while insisting to your long-suffering wife that it was all-important and necessary research? Was it stimulating? Was it excessive? Was it real?

When I first got on the r/NoFap sub, I looked down on these folks. I thought it was ridiculous to spend half your day on Pornhub masturbating to videos, and I thought it was pitiful to then spend the other half worrying about it on Reddit. I thought the whole project was undermined by its being carried out online. They needed to take a walk outside, talk to other human beings in person. And part of me still feels that way.

Here's the thing, though. It's not like you can't meaningfully communicate online. In fact, while the intellectual rigor on that thread left a lot to be desired, they *were* engaged in a kind of ideal dialectic: direct philosophical engagement of fully disembodied brains. It was weird to me, and it was plainly weird to them. It wasn't exactly what they'd signed up for. But it was eye-opening too, watching it all happen. The people on that thread couldn't say whether the folks they were interacting with had long hair or short, whether they were skinny or fat, whether they had halitosis or psoriasis or looked often at the ground, whether they were young or old, White or Black, men or women. They were forced to evaluate and respond to one another purely on the strength of their ideas and the eloquence of their expression, and in a world where the human body has often been the basis of shame and disenfranchisement, there's beauty in that.

I was missing the point. The conversation wasn't about masturbation. And it wasn't even about separating from the internet. That's not a possibility anymore. It was about distinguishing what the internet can provide from what it can't, and the subreddit itself was an example of the former.

It took a while for that discussion to wind down, and it was late by the time I finished sketching out my notes on it. I went upstairs, and the bedroom was dark. My wife was asleep. She rolled over as I lay down and threw her arm across my chest without waking. It's a usual thing, but the time on Reddit gave it an air of surreality. Who was this body? When had we last spoken? What had we said? Was it real or just pixel? How can you tell?

> Good fucking night, r/NoFap. Here's to a better tomorrow.
> u/uzumati

SEMEN ISN'T PRODUCED IN THE SPINAL CORD, AS DA Vinci illustrated around 1492 in his drawing of a hemisected man and woman in the act of coition. It does not come from between the backbone and the rib cage. It is not produced, like Alessandro Benedetti thought, in the kidneys. It's not a residue of blood, which only the male, with his adequate vital heat, can transform. Neither is it made, as Aristotle wrote, around the eyes; dark circles beneath the eyes and vision problems are not the hallmark of an incorrigible masturbator, and crying does not deplete your masculinity. It doesn't come from the brain, per the ancient manuals of China, or from any location in the nervous system, and it cannot be circulated, as the Taoists maintained, via this system back to the brain.

Despite Giles of Rome, it can't shape menstrual blood into human form the way an artist carves idols from wood. Neither, as Ibn Sina preferred, does it create life from menses as from mere milk the rennet creates cheese. You can't combine it with hair to produce the Paracelsian homunculus, and it can't be read like tea leaves or the flights of birds or worn like an amulet against the depredations of evil spirits. One drop of it is not derived by the bodily distillation of sixty to a hundred drops of blood, and contrary to the teachings of Muhammad, it isn't a clot. Swallowed, it cannot burst through the forehead in the form of the god Thoth or the goddess Athena. Cast into the sea, it doesn't produce a foam from which Aphrodite rises. It cannot materialize in the womb or Fallopian tubes via divine action. It has no magical powers. It cannot carry the soul. Oaths sworn upon it are not consecrated.

Its color does not correlate with the skin pigmentation of its ejaculator. It does not inherit the flavors of coffee, anise seed, onion, or lemonade. It isn't possible to conclude that selenium, L-carnitine, or coenzyme Q10 drastically alters its production, and its volume cannot be increased to any notable degree by the ingestion of internet-marketed supplements. It is not greatly affected by Mountain Dew.

Neither is its quality degraded by frequent masturbation. If you coax it into existence alone in your stepmom's house while watching *Golden Girls*, no god minds. Its emission while on the internet does not induce erectile dysfunction. Its emission while on the internet is not evidence of a Jewish conspiracy to drain young men of their vigor and purpose. Its emission while on the internet does not cause blindness or depression or brain fog or destroy productivity or decrease spiritual awareness, and its retention will not boost confidence or prevent hair loss or strengthen the immune system or give you the emotional stamina to take meaningful and rewarding risks. The desire to abstain from masturbation does not correlate with mean masturbation frequency, maximum number of orgasms (weekly), onset of masturbation, or internet pornography consumption. The desire to abstain from masturbation does correlate with the feeling that your life is not within your control.

u/always122 said something about their fellow redditors on the r/NoFap sub during that conversation, as the chat was being torn apart by centrifugal forces seemingly beyond its control. "We are a lost people," said u/always122. When I first witnessed the thread unfolding, I thought they were lost, too. They'd left behind many of those daily interactions with the physical world—school, work, recreation—that had previously moored people to the flesh-and-blood fact of other human beings. And in place of those interactions, they had entrusted themselves to a net of emojis and videos and algorithms. I still think they're a lost people. Now I just don't know who isn't or if that's a bad thing.

If two people should meet someday on the r/NoFap sub, if one should be charmed by the other's witticisms, if they should upvote each other, strike up a correspondence, DM, text, follow, incubate a relationship entirely via pixels, they could actually get to know each other pretty well. If they so chose, they could even meet on a virtual mountaintop shrouded in mist. One could be a blue ogre or a daring pixie, the other a knight or an elf or a gargantuan teddy bear. They could take a walk through a lava-strewn plain, hold hands, barter skins, do their taxes. They could talk about the future too, and that future might soon entail strapping on their virtual reality crotchsets, reclining beneath a bower of binary magnolias, and frantically rutting each other's avatars. Is that healthy? Is that good? Is that the felt reality of love? At one time I would have said no, but I'm no longer so certain. Who am I to say where orgasms belong?

V

Menses

When you lay your body in the body
entered as if skin and bone were public places

Claudia Rankine, *Citizen: An American Lyric*

Terms

n. Aunt Flow, the flow, shark week, red badge of courage, moon time, that time of the month (TOM), the tide, the crimson tide, girl flu, Carrie, the curse, code red, catamenia, the business, lady business, none of your business, period

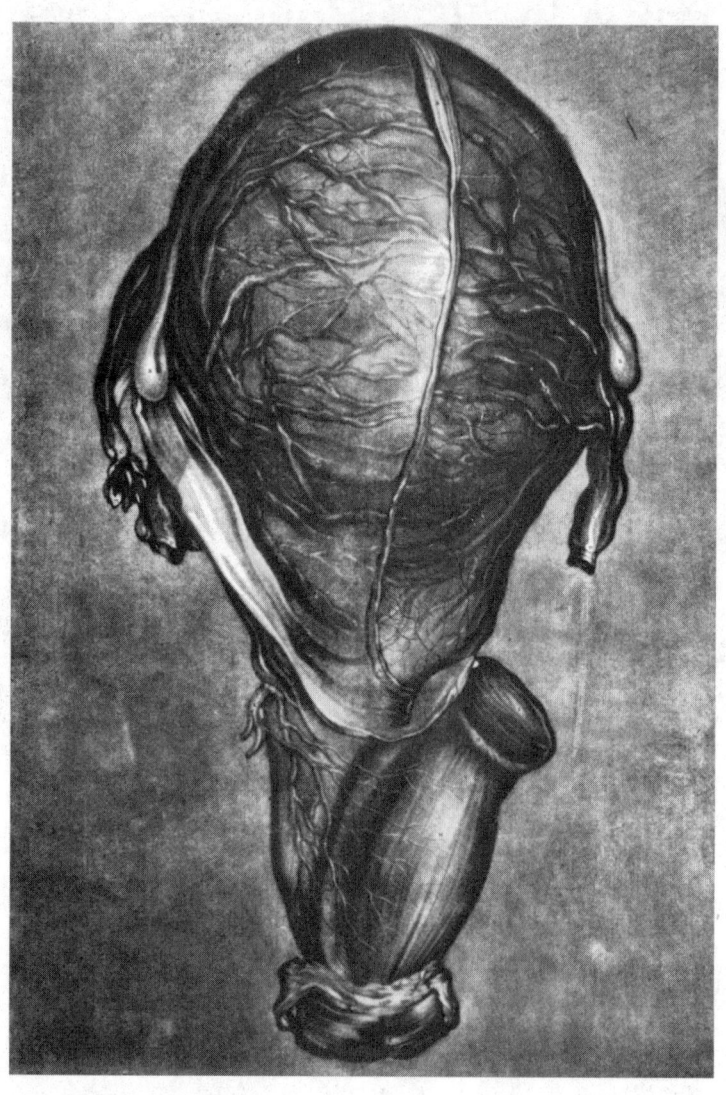

Illustration of a uterus, Charles Nicolas Jenty, 1757

Biological Prologue: The Spontaneously Decidualizing Mystery

For thirty or forty years, starting at the oily onset of puberty and ending in the electric thrum of menopause, a woman's uterus and ovaries go through a roughly twenty-eight-day pas de deux that culminates in either pregnancy or menstruation. It's a heavily scripted dance, choreographed in large part by a little pea of glandular tissue and axonal projections snuggled into the middle of the head, about two inches behind the eyes. This is the pituitary gland. According to some, it's a third eye, like the pineal, partially atrophied in our evolutionary history, which when opened pierces the veil that separates this realm from the next and transports one on cosmic voyages of enlightenment, but by traditional observational measures, it neither travels nor sees. It sits hunkered down under the brain, in a little saddle of bone called the sella turcica. It's accessible to those who really want to open it, but only via some serious endoscopic finagling up the nostril.

At some point in the life of an unsuspecting human, often between the ages of ten and fifteen, on an arbitrary day of the pituitary's choosing, the gland up and releases a substance called follicle stimulating hormone, FSH. This FSH enters the bloodstream, circulating throughout the body until it reaches the ovaries. Since birth, these two organs have largely slumbered in the abdomen, each with its clutch of eggs several hundred thousand strong, every one of which is nestled in a tiny fluid-filled cyst called a follicle. But at the arrival of FSH, the ovaries gently stir, a follicle swells, and after about two weeks, an egg pops out and coasts down the Fallopian tubes toward the uterus.

The egg gets a great deal of attention in this process, and sure, it's important (life, etc.), but the ovary has a partner in all this, and from the perspective of energetic expenditures, the departure of the egg from the ovary is just so much sleight of hand. In terms of where the body is really putting itself out, the real action is going on down in the uterus.

The uterus isn't much to look at, just a little lump of muscle about six inches below the belly button. Other than the bladder, which it sits on top of, its main company down there is the rectum. It's shaped like an upside-down hollow pear, and its inner surface is lined with a thin layer of tissue called the endometrium. There's nothing remarkable about the endometrium at first glance, but something starts to happen here while the follicle is growing. Between days four and seven of the cycle, with the arrival of FSH and the increasing tide of follicular estrogen, the endometrium begins to thicken. This thickening is the creation of a layer of spongy tissue where a fertilized egg can land and grow, but we're not just talking about putting down some padding and nosh. Should an egg get fertilized up in the Fallopians and should the resulting blastocyst touch down here on the endometrium, this will become the site of one of the most precarious and least understood events in human biology, the tenuous détente wherein a woman's body partially suppresses its own immune response and allows a foreign body to physically invade its tissues. It's not a small thing to let a group of unknown cells just grow inside you. Anyone who's had diarrhea knows how the body usually reacts to these kinds of incursions, and anyone who's had cancer knows why the body usually reacts that way. To host that blastocyst and not die requires some serious site prep.

The spongy tissue is actually riddled with glands, about fifteen tucked into every square millimeter. Cells are rapidly dividing and proliferating here during this time, and as the endometrium develops, getting thicker and thicker, these glands become deeper and increasingly convoluted and tortuous in shape. Something has to feed the whole apparatus, so growing between the glands is a network of minuscule spiraling arteries. In humans and a few other species, the endometrium also goes through an additional process before the arrival of the egg: spontaneous decidualization. In decidualization, the endometrium thickens and develops. Some endometrial cells metamorphose from elongated structural cells to rounder cells capable of providing nutrition, while cells on the very surface of the endometrium develop little flowerlike projections called pinopods where the blastocyst will attach. The whole cellular matrix of the endometrium becomes denser during spontaneous decidualization, and it's flooded with immune cells like uterine natural killer cells and macrophages, which form in the bone marrow and home in on the uterus through a process not yet understood.

What's being created, in PhD-speak, is a columnar epithelium formed into tubular glands, differentiated into secretory and ciliated cells, fed by a highly vascularized stroma, and seeded with its own highly specialized immune agents. It's not just a spongy lining. What women's bodies create each month is an entirely new organ.

Now, should the egg go unfertilized or fail to implant successfully in the endometrium, the secretory phase begins. The uterine muscles contract, the endometrial blood vessels constrict, and the entire organ the uterus just built begins to break down. Little by little, it gets shed, exiting the body through the vagina. It's called menstruation or, more commonly, a period.

This isn't news. In a general way, everybody, with the exception of small children and a few soon-to-be-startled ten-year-olds, knows what a period is. But those ten-year-olds are right to be startled. A period *is* crazy. It's like growing and throwing away a liver every single month. And it's strange then, considering how involved menstruation is, considering how many of us on this planet are menstruating right now, that we don't know why it happens.

What they don't teach you in any sixth grade class is how rare the menstrual cycle is. Hyenas don't menstruate, for instance. Neither do cats, elephants, Weddell seals, or the three-toed sloths of Barro Colorado Island, in the middle of the Panama Canal. Dogs, contrary to popular belief, don't menstruate. (They occasionally bleed *before* their fertile phase, due to extravasation during the growth of the uterine lining, but it's definitely not menstruation.) All these animals experience a similar but very different cycle, the estrus cycle, where their uteruses generate a rudimentary endometrium to receive a fertilized egg. But it's a basic lining, and there's no spontaneous decidualization. If no egg implants, the endometrium is simply reabsorbed by the animal. No loss of tissue, no loss of energy: great system. In fact, it's such a good system that just about every mammal out there *doesn't* menstruate. The whole extravagant rigmarole is limited to a select handful of species. A few bats, a kind of shrew, a single rodent, and the primates, including us: menstruation is all ours.

And it just doesn't make any sense. It takes a lot to grow all those glands and arteries and build up a whole mini–immune system, and the body isn't ordinarily so spendthrift with its energy. It doesn't just grow and toss out organs willy-nilly. It must have a very good reason to do so. We just don't know what that reason is.

The Limit

At the touch of a menstruating woman, bees flee the hive, mares in foal miscarry, boiling linen turns black, imperial purple fades, bread refuses to rise, the razor's edge becomes dull, copper rusts and emits a foul smell, wine spoils, ham spoils, pickles spoil, mirrors fade, and sores develop on the penises of men.

The menstruating woman frightens the prey of the Eskimos and turns the blood of the Mae Enga black. If she walks naked through a farmer's field, the caterpillars, worms, beetles, and other vermin fall to the ground, deceased. If the walk occurs at sunrise, the crop dries up, the young vines shrivel and die.

Drawn by her heady and supernatural essence, ghosts flock

to her with such enthusiasm that she must bury her menstrual clothes to prevent their use by wayward spirits. If her spouse goes to work on a tomb during her menstrual period, immortality eludes forever the man interred there.

The menses itself is venomous. It emits a smell or vapor or fume or ray, and it is the source of menotoxin, a poison that circulates throughout the body, appearing in the sweat, the milk, the blood in the veins. Threatened with this substance, hailstorms settle, whirlwinds dissipate. Men become weak and old. Frogs exposed to menotoxin and given narcotics get high, higher than those given narcotics alone. Rodents injected with it become disoriented and die. It damages mums, anemones, roses, peas. Lupines are unaffected, but cowslips invariably wilt.

Though it can absorb and purge a man's gonorrheal sores, to touch the penis to menses in fornication brings disease and death. Children conceived in the fluxus emerge deformed, with epilepsy and leprosy, while infants fed upon menotoxic breast milk develop cholic and asthma.

A special danger: occasionally menses comes out of the nose.

In France, the menstruating woman is not to gather the yellow trumpets of the chanterelles in June or pick the delicate clusters of the cabernet franc grape in September. Nor is she to refine sugar or care for silkworms. In Britain, she is forbidden to cure a ham, which requires the steadfast rubbing of the pig's raw thigh with salt and spices. "Whilst in some women menstruation was a comparatively local process," says the *British Journal of Medicine*, "in others it was accompanied by characteristic emanations from the rest of the body, and these, proceeding from the hands and reaching the pork in the process of rubbing, could hardly fail to contaminate it." She is, in other words, "unfit to manipulate ham." In Shinto, the indigenous religion of Japan, she is not allowed to visit the tops of sacred mountains. For the Beng, it is important that she not go into the forest or caress a corpse or tend a fire. The food she cooks is not to be eaten by either her husband or the Masters of the Earth.

In Nepal, she must seclude herself in a small hut. In Thailand, she cannot circumambulate the stupas. At the beach, she must not swim, lest she draw in sharks. And in the Orthodox Jewish community, she must track her period using a small cloth known as a bedikah, which, should she experience irregularity, must be inspected by a rabbi, either in person or via the Tahor app. In Texas, any physician evaluating a pregnant woman is legally required to make note in her medical record of "the amount of time that has elapsed from the first day of a woman's last menstrual period," while in Missouri, the practice of tracking women's periods has been modernized. They use a spreadsheet.

THE PATIENT ARRIVES AT THE EMERGENCY ROOM IN the early afternoon complaining of abdominal discomfort. Her vital signs are relatively normal. Her blood pressure is low, but her blood pressure is usually low. The pain is most acute just below her navel and to the right, where her waistband usually sits, and it has gotten worse even since her arrival at the ward an hour earlier. She's dizzy, as well, but besides this, she has no other symptoms.

She takes no medications. She has no allergies. She doesn't smoke, she drinks moderately, and she is physically active. She hikes and bicycles a great deal. She is thirty-two years old, unmarried, with a one-year-old daughter, and she runs her own business, a small café behind the courthouse. She is anxious to have the issue resolved so she can return to work. She has driven herself to the ER.

There's a chance it could be appendicitis. A blood test would verify her white blood cell count. Urinalysis, meanwhile, would rule out a kidney stone or urinary tract infection. She receives neither. Instead, she's dressed in a hospital gown and led to a seat behind a curtain, and that's where she sits, and she is still sitting there another hour later when she begins to slip in and out of consciousness. Time passes. It's impossible to say how much, and finally, a nurse, a janitor, someone passing down the corridor by chance, notices there is a woman sitting in a chair and she does not appear to be entirely alive. A doctor is summoned then. He rushes down the hall, looks at the patient and then at her chart, and it is thanks to this doctor—Dr. Yasui, he deserves to be remembered here—that the woman shortly finds herself being rushed to surgery.

What Dr. Yasui has recognized are the symptoms of an ectopic pregnancy. The condition, where a fertilized egg implants itself somewhere other than the uterine wall, is hardly uncommon—estimates suggest around one in forty pregnancies are ectopic—though, as with many pregnancies in the first twelve weeks, women often don't even realize anything has occurred. The misguided blastocyst is sloughed off along with the endometrium during menstruation. But sometimes an ectopic pregnancy can be fatal. In this patient's case, the egg has implanted itself in one of her Fallopian tubes, and its growth has ruptured the narrow passage, causing her to hemorrhage internally. She is losing blood, in other words, a great deal of it. Within minutes of the encounter, she will be prepped and wheeled into the surgery, and there she will receive the medical procedure that will save her life, an abortion.

She'll go on to open a second restaurant, to give birth to a second child, to continue hiking and biking, and to spend the next decades feeling extremely thankful for Dr. Yasui and for that operation and for life in general. But at that moment, as the doctor, struggling to keep her alive, rolls her in the direction of the surgery, at that moment she is just another woman quietly bleeding to death in a hospital hallway. There is nothing unique about her except that she is, or rather she will be, my mother.

IF YOU'RE LUCKY ENOUGH TO HAVE A MOM, YOU DON'T really get to know her for a while. By the time you have a consciousness, a memory, she is simply *there*. For the first years of your life, you and she exist in the ever-living present, and for all you know, this is how it has always been. You love her deeply, you resent her regularly, and sometimes when the world begins to conspire against you, you put your head in her lap and cry. For a long time, it never really crosses your mind that your mother might have existed before you, and it's weird then when you start to learn about this earlier woman, this woman who was not yet your mother.

You build her out of bits and pieces, little stories she mentions while paddling in a canoe or anecdotes she slips into conversation at the dinner table—that she'd majored in philosophy at Penn State but then moved to New York City to become a nurse, that she'd abandoned nursing school to return to Pennsylvania and open a restaurant. And it's hard to know what to do with all this information. A lot of it comes before you're prepared to understand it, which is part of the reason little kids ask so many questions. What was I to make, for instance, of the knowledge that my mother once attended a bra-burning in State College or that she was nearly incinerated as a girl when her petticoat caught fire? What are petticoats? What are bras? Why is women's underwear so burnable?

The first time I remember my mom telling me about Dr. Yasui and the abortion, I was eight. Death was on my mind then—my best friend's mom was dying—and maybe it was after I attended that funeral, my first, that it struck me how close my mother had come to death. It's a hard thing for a kid to wrap his mind around, not just the loss of the person most central to your existence but also the idea of that alternate reality in which you don't exist.

But I never much thought about the abortion part of things. I was just glad my mom was still alive.

THE CONFUSING THING ABOUT MENSTRUATION IS that while it has not developed widely, it was so advantageous in certain circumstances that it developed independently in vastly different mammal lineages *at least four separate times*. This process, the same trait evolving of its own accord in different species, isn't too uncommon. It's called convergent evolution, and it's why the euphorbias of Africa and the cacti of North America both have fleshy accordion-like stems for water storage. Faced with the same strong, specific evolutionary pressure—a scarcity of water—they independently struck upon the same solution.

But what's our scarcity of water? The evolutionary pressure must have been significant to offset the obvious energetic disadvantages of creating and excreting an organ every month, and it must have been very specific, since menstruation didn't evolve more widely. And the adaptation must have been uniquely effective, since it developed again and again (and again and again) in different species. But what was it? It's hard to see what the menstruating species all have in common. They eat different foods, live in different climates, construct different societies. It's a mystery. An elephant shrew, an orangutan, a mastiff bat, a spiny mouse, and a human all walk into a bar. When they walk out, they all have their period. What happened in the bar? Why does menstruation exist at all? The answer must lie somewhere in those animals' pasts, but for now, as was noted recently in the

American Journal of Obstetrics & Gynecology, "the most honest and shortest answer to this question is 'we do not know.'"

One of the reasons we know so little, of course, is that scientific studies long focused on men for fear that female results would be skewed by the fluctuating hormones of the menstrual cycles. But on the more practical side, we don't know much about periods because we learn a lot of what we know about humans by running experiments on our most convenient biological analogs, pigs and rodents. But the pig doesn't menstruate, and until the recent discovery of a period in the Cairo spiny mouse, neither did any rodents appear to.

The one thing we do sort-of-know is that evolution doesn't seem to have selected for menstruation at all. Instead, it's believed menstruation is just a side-effect of the actual evolutionary adaptation: spontaneous decidualization.

In most mammals, the endometrium only decidualizes in response to chemical signals sent out by a fertilized egg. But the one thing all menstruating mammals have in common is that their endometria decidualize spontaneously, whether the egg is fertilized or not. In fact, they do it before a fertilized egg could even be present. So the real question isn't actually why do we menstruate, it's why do we spontaneously decidualize? Why this massive undertaking when it's not even clear whether a fertilized egg is coming down the pike? Why not be like the cow and the goat and the yak and the orca and virtually every other mammal and build up the endometrium only when it's actually needed? The answer's not clear, and it probably won't be for several hundred generations of Cairo spiny mice.

IT WASN'T UNTIL A LITTLE WHILE AFTER I FOUND OUT about my mom's abortion that I realized abortions were a thing people cared about. It was the twentieth anniversary of *Roe v. Wade*, though I didn't know what that was at the time, and there was going to be a Pro-Life rally on the strip mall in our Pennsylvania town. In the days beforehand, some of the girls at school discussed the signs they'd be bringing, talked about what they were wearing, made plans to get rides with one another or to meet beforehand in the Long John Silver's parking lot. I didn't really have much interest in going. It was supposed to be on Saturday morning, right in the middle of my prime cartoon-watching hours. But my friend Jesse dragged me along. He and his parents were going to counterprotest. My mom didn't go. Saturday was the busiest day at her restaurant. "Causes are for people without jobs," she said.

It was hard to understand, as we got closer to the center of the rally, exactly where all the people had come from. There seemed to be far more than lived in town. They'd gathered in the parking lots at Kmart, Joann Fabrics, Blockbuster Video, and then lined the sidewalks along the strip. Some were chanting and singing hymns and strolling. Others wore T-shirts with slogans—ABORTION STOPS A BEATING HEART—and carried signs. Mainly they just milled up and down the street, waving to cars.

A few other families had come along for the counterprotest. There were maybe ten of us. I don't know what our goal was. It cannot have been, I think, to win hearts and minds. I suppose the adults just felt that someone should represent the other side of the matter. So we circulated. The strip was crowded with people, and to make our way around we often had to ease ourselves and our signs between those who were there for the rally. They parted way for us, assuming at first that we too were there to support the Pro-Life cause and only realizing when we were already among them exactly what we were about.

Reactions were mixed. There was some vitriol, some talk of murder and shame, but many were merely polite. A choir, I remember, broke into a chorus of "We Shall Overcome" as we passed. The reactions to Jesse and to me were more subdued. They looked upon us sadly. No one said a word to us. Though these adults would sometimes argue with the adults on our side, they made no attempt to engage us nine-year-olds in discussion or criticize our beliefs. They seemed almost afraid to interact with us. We were, after all, not their children.

AROUND THE TIME I LEARNED ABOUT MY MOTHER'S abortion, my friend Jake's mom was dying of cancer. Years before, when she was pregnant with Jake's younger brother, her doctor had discovered a mass in one of her breasts. They could have treated it, but doing so would have required ending the pregnancy. She chose to delay treatment and take the pregnancy to term, and she gave birth some months later to a little boy. But the cancer had progressed in that span, as it continued to do while Jake and I were growing up. By the time we were in elementary school, there wasn't much more the doctors could do.

Jake never spoke of his mom's illness, and it never crossed my mind to wonder why his dad had started picking him up from my house instead of his mom. At some point, I found out she was sick. A teacher at school took me aside one day. And a little later, I went to the funeral (I spent most of it in the parking lot, too nauseous to speak). But it wasn't until much later that I learned any of the details about getting the diagnosis while pregnant. I doubt anyone told Jake or his brother. That's a heavy trip to lay on any little kid, letting him know his mom died so her child could live. I'm not sure if Jake even knows now.

I went over to his house after school once around that time. Jake ran through the living room without stopping and disappeared up the stairs, talking over his shoulder about some new game he had in his room. Jake was always running somewhere, always talking over his shoulder about something. At that time, his brain was like one of those prize wheels at the fair. Whenever it began to slow, he sent it spinning again.

We'd been playing a while upstairs when we heard the front door open. Jake began to talk more loudly. He wanted me to see his zebra collection. His favorite animal was the zebra. He had a whole box of zebras. He wanted to show me his skateboard. He'd been practicing in the alley, check out this scab, did I want to see a kickflip? We heard his father's muffled voice through the floor, then a pause, then footsteps, then the voice echoed clearly up the stairs.

"Jake, did you say hi to your mom?"

"What?"

"Come down here."

"Cutter's here."

"Then he can say hi, too."

Jake rolled his eyes, spun the wheels on the skateboard, and let it fall to the floor. He left the room without a word, and I followed.

There was a set of pocket doors that divided the living room from the dining room. I hadn't noticed they were closed when we ran past, but they were open now. The table was gone from the dining room, and someone had pulled the curtains. It was around dusk. The room was sunk in a greenish hue, as though the whole home were underwater. In place of the table, there was a hospital bed, and in the bed was a body. The skin was pale and waxy, and the hair lay in wisps across the scalp. A hand, bony and distended like the hand of an alien, protruded from the covers, and it was into this hand that Jake, dropping into a chair, thrust his own. I don't remember much what we said. I remember a bag hung from an IV stand. You could hear it dripping.

THERE ARE TWO BIG HYPOTHESES AS TO WHY WE spontaneously decidualize. The first has to do with the uniquely invasive nature of our fetuses.

In all mammals, the fertilized egg is essentially parasitic. It feeds upon its host to grow. But in humans and other menstruating mammals, it is more like a cancer. The blastocyst lands on the uppermost layer of the endometrium, settling on a spot between the glands and sticking to the pinopods there. Like some sort of drilling apparatus, its internal machinery, the inner cell mass, swivels within the cell until it directly faces the endometrial surface. Then, covering itself in a layer that will hide it from the mother's immune system, the blastocyst penetrates between the mother's cells, sinking into and being encapsulated by the endometrium. Implanted there, nourished by glandular secretions, it grows, quickly eroding through the walls of neighboring glands and sending out specialized cells—extravillous trophoblasts—that invade and replace the gland walls, bathing it in nutrients. These trophoblasts then continue burrowing out and down into the endometrium, anchoring the blastocyst and eventually reaching the mother's spiral arteries. There the blastocyst destroys and remodels the arterial walls, replacing the mother's muscle cells with a flaccid structure formed of its own cells and thus ensuring a continuous low-resistance supply of blood.

Given the opportunity, the blastocyst doesn't stop there. After a Caesarean, for example, the regeneration of the decidua is often impaired at the site of the uterine scar, and sometimes in such circumstances, those specialized cells, the extravillous trophoblasts, grow down through to the lowest levels of the endometrium. They'll occasionally reach the smooth muscle of the uterus and grow into it. They can even grow right through that muscle and begin, like a tumor, invading the bladder and rectum. These are all life-threatening conditions—placenta accreta, increta, and percreta, respectively.

The first theory of why we develop the decidua ahead of time is that along with nourishing the blastocyst, it's also very dense. It prevents the drilling from going too deep. It's a little like putting dinner on the table while also locking the door to the kitchen. In this telling, there's a fetal-maternal evolutionary arms race, ever more invasive fetuses demanding ever more prepared uteruses, and the eventual result is spontaneous decidualization and the period.

It gives you more respect for the act of pregnancy, and it helps to explain why every human pregnancy, whether we know the exact chances or not, is itself a life-threatening condition. The comparison to cancer isn't just a handy metaphor here. It's believed that part of the reason human cancers are so invasive compared to those of animals like horses is that our cancer cells reactivate the kinds of gene expression present in our embryos. Human cancer is a kind of gestation, or to think of it another way, in our species, every gestation is basically controlled cancer.

This is scary stuff. My wife had placenta accreta, though no one realized until our daughter was born. The placenta just refused to come out of the uterus. The doctors took turns tugging on the thing for the better part of an hour while she dry-heaved into a little bucket. It's not like the placenta just gets stuck to the uterine wall in a situation like that. The arteries and veins of one have grown into the other. They were trying to physically tear out part of her circulatory system. It was gruesome. You're lucky in that situation if you don't lose the uterus. We felt lucky just that she survived. Not that long ago, she wouldn't have. The doctors looked terrified. I must have too. A nurse suddenly noticed me watching and yelled at me to sit down. Fathers, it turns out, often faint during delivery.

IT WAS HARD GROWING UP ALONGSIDE JAKE AFTER his mom died. He careened from one extreme to the next. Zebras were out, chess was in, his family was out, being straight edge was in, straight edge was out, weed was in. There wasn't much saying who he would be the next time you saw him. One day he wouldn't touch a can of Pepsi because of the caffeine, the next he was talking about beating up the dealer who'd sold him an eighth of shwag and stems.

One day, in home ec class, a few of us had finished our sewing projects and Jake whipped out his chessboard and started playing with a kid who'd just transferred to our school. This was in the heyday of Yo Mama jokes, and as Jake was tearing this new kid apart on the chessboard, the kid just kept laying these lines on him.

"Yo mama so stupid when she hear it's chilly out she grab a bowl."

"Yo mama so poor she go shopping on trash day."

"Yo mama so fat she on both sides of the family."

And finally Jake, without looking up from the board, said very quietly, "Yo, that's not funny, my mom died of cancer."

And the kid laughed, and then looking around at the rest of

us, he laughed a little more quietly, and then he was like, "Yo, really?"

"Yes, really, my mom is dead."

"For real?" The new kid looked to the rest of us. We nodded. "Shit, man," he said. "Shit."

Then Jake smiled. "I'm just messing with you. My mom's alive."

It was hard to watch, and it's still hard for me to explain what happened next. The other kid got angry and started telling Jake how messed up it was to say a thing like that, and then Jake flipped back again and started claiming his mom was dead, but the other kid wasn't buying it this time. He was still tearing into Jake about pretending his mom was dead, and Jake just kept saying "No, no, she's really dead. I swear she's dead." Almost like he couldn't stop saying it. Though he couldn't help but smile and kind of laugh. And finally the kid looked around at all of us again. We were frozen. This was the most excruciating thing that had ever occurred. No school can teach you how to act in a situation like that. And I don't know if the new kid realized Jake's mom was dead. I think he did. But he couldn't stop being angry and he got up and just walked out of the room.

"Fuck this school," he said.

THERE'S A SECOND HYPOTHESIS ABOUT WHY SPON-taneous decidualization exists, and it builds upon the first. It takes into account that the gestation of menstruating mammals is inordinately dangerous, and it also factors in that it's comparatively long. The mouse living under your fridge will make a litter in under three weeks. The Cairo spiny mouse requires as much as six. This hypothesis is based on the recent discovery that decidualized cells fulfill one other function. Though we don't yet know how, they also assess the viability of the developing blastocyst. If they detect something amiss, the whole operation changes gears. They inhibit the release of the cytokines and growth factors necessary for the embryo to develop, and the whole structure of support instead begins to disintegrate.

In this understanding, life is short, pregnancy long and dangerous, and so a woman's body is equipped with a fail-safe. Decidualization isn't just a way of feeding and fending off a fetus. It also endows the endometrium with agency. It gives her the power not only to create life but also to select against it.

So what is a period then? It's not just the shedding of the uterine lining. It's not a nuisance or an inconvenience or the expunging of the sin of *Brahmahatya*. It's the decidualized endometrium exercising its monthly right of disintegration.

THE POWER OF MENSTRUAL BLOOD MAY BE HARnessed, says Pliny. A plague of locusts can be eradicated, for instance, if a menstruating woman but walks through the field with her clothes pulled up above her buttocks. It may even be sufficient to simply walk barefoot with her girdle loose and hair disheveled. The ancient midwives Salpe and Lais counsel that malaria and rabies are "cured by menstrual fluid on wool from a black ram enclosed in a silver bracelet," and the first century obstetrician Sotira advises awakening epileptics by rubbing their feet with menses, noting it is "much more effective if it is done by the woman herself without the patient's knowledge." Lepers are cured by washing in it, says Hildegard von Bingen. We should procure as much of the blood as we can.

And why not? The endometrium is the closest thing we know to a phoenix. This organ perishes each month, and no sooner has it gone than it begins to grow again. It beats back cancer. It heals without scars. It is not full of menotoxin but of endometrial stem cells, and those cells are capable of things that could easily be mistaken for miracles. Transplanted into atrophied muscle fibers, they promote the formation of new blood vessels. Grafted into dead tissue following a heart attack, they transform into striated cardiac muscle cells, and injected into the bloodstream, they reduce the size of tumors.

You start to understand why some cultures revere menstruating women. I remember a friend telling me about this Cherokee legend where a stone-skinned monster, Nun'Yunu'Wi, comes down from the mountains in search of human flesh. He is shaped like a pale old man—I imagine him like a congressman—and he is carrying this magical cane. He sticks the cane out, then sniffs the tip of it, and that's how he can tell where people are. So he comes toward the village, and the warriors go out to fight him. They throw everything they've got at him, arrows, spears, rocks, the works, but it all just bounces off, so they beat it home while he keeps on coming. Nothing can stop him. They're all doomed. But when he gets to the village, the first person he comes to is a menstruating woman. He runs from her in terror, a dribble of blood at the edge of his mouth. The next person he meets is another menstruating woman, and the same thing happens. Seven times, the monster crosses the path of a menstruating woman, each time growing weaker and weaker, until finally, blood pouring from his mouth, he collapses. They burn his body. Inside is a magic stone and the lumps of red paint that will become the emblem of the Cherokee people.

A FEW YEARS AFTER HIS MOM'S DEATH, WHEN JAKE and I had long since drifted apart, I ended up at his house one night for dinner. This must have been middle school, maybe high school. The place was clean and kempt and bright. His father had remarried by then, and he walked the halls in a robe, playing the banjo, while Jake's stepmom cooked. We had chicken for dinner, peas. For dessert, there was applesauce. No hospital bed occupied the dining room. No woman wept there, and no boy looked out the window and asked if he could go now. It was a room like any other, but we ate in the kitchen.

Jake was fucked up by the experience. His whole family was. Deeply, and for a long time. And because of that I thought I was lucky that Dr. Yasui was there that day—which was true— and I also thought my mom was smarter—which was not.

It just didn't make sense to me what Jake's mom did. Why not get the abortion? She could have had more kids later, she could have lived, she could have been there for her children. I respected what Jake's mom had done—it was a breathtaking sacrifice—but that's different from understanding it. And because of that, I spent a long time thinking there was this huge void between our mothers, which wasn't totally wrong: Jake's mom took one path and my mom took another. But that's missing the point entirely. We're all missing the point. Jake's mom faced a choice, one so personal, so literally internal, that no one could ever make it for her. I don't need to agree with or even understand her choice to be moved by the degree to which it was her own.

JUST TO WITNESS MENSES IS DANGEROUS. THE MAterial propagates itself through the faculty of vision. Should one woman glimpse the menses of another, she herself will begin menstruating, and should that menses be glimpsed by two others, then four, then eight, in an exponentially expanding network of menstrual proliferation . . . the imagination reels. As Pliny says of the subject, "There is no limit to woman's power."

VI

Milk

Macbeth
Act 1, Scene 5
Inverness, Macbeth's Castle

LADY MACBETH
Come to my woman's breasts,
And take my milk for gall, you murdering ministers,
Wherever in your sightless substances
You wait on nature's mischief.

Terms

n. numnum, milk bar, milk truck
v. nursing, suckling, chowing down, hitting the tit

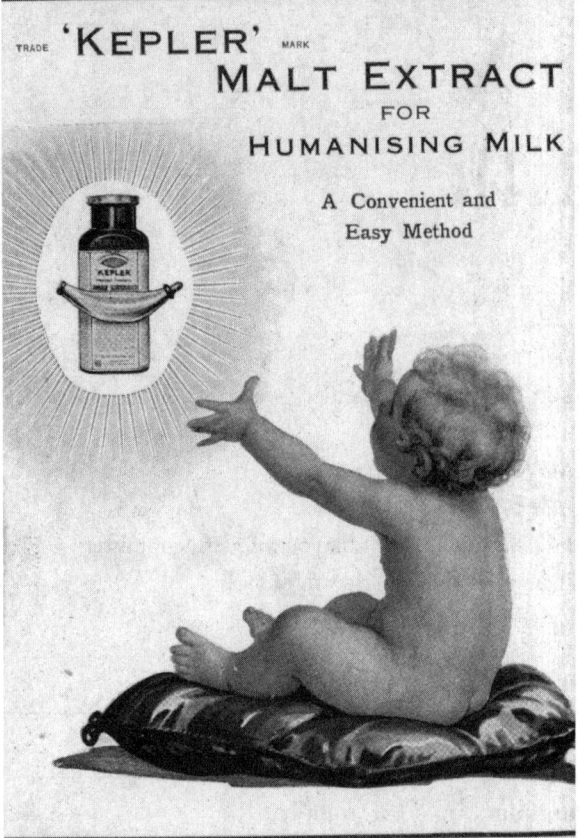

Advertisement for Kepler Malt Extract infant formula, 1890–1920

Biological Prologue: First Milk

Out in the ether, a circle forms. Your retinas aren't yet developed enough to process most color, so it is a gray circle. Likewise, your field of focus is limited to a distance between eight and ten inches. A gray and fuzzy circle. With little control of your muscles, you can only spectate as it grows larger, blotting out more and more of your field of vision. You watch as another slightly darker circle forms within it, and as these two circles swell, you see yet another, yet darker circle, form at their center, until finally, a sort of gauzy bull's-eye hovers before your eyes.

A smell comes to you now. Your nose, though sensitive, lacks the frames of reference that would allow you to recognize this as the sebaceous secretions of the Montgomery glands, but the odor—pungent, oily—sockets into your brain as a key within a lock. Coaxed into being by this scent, lightning crisscrosses the brain matter at the base of your skull before scattering out into your body. Without you willing it, you inhale deeply, craning your neck and searching the air with your lips, and at this moment, the darkest innermost circle breaks the focal plane. Soft, rounded, shaped something like a raspberry, its surface dimpled with pores, it is a nipple. A few beads of colostrum tremble on its surface. You tremble too, and a breast falls on your face.

The blocking out of light, the brushing of the nipple against your cheek, and the sudden intensification of scent all work together to initiate a series of coordinated reflexive movements in your body. First, your head turns toward the nipple, nuzzling side to side, the rooting reflex. Then, as the nipple brushes your upper lip, your mouth puckers and grasps, drawing it in, the snout reflex. Finally, as the nipple touches the roof of the mouth, the opening to the nasal passage closes and the muscles in your jaw contract rhythmically, the sucking reflex. The mouth sucks and sucks, and here, in your toothless, high-arched palate, the breast discovers a gorgeous correspondence. With each suck, the areola is pulled farther in until the oral cavity is filled with a good two inches of flesh, suctioned à la a Sugar Daddy against the roof of the mouth, with the nipple positioned at the top of the throat like a nozzle.

There is nothing you can do about it. This is all the work of the brainstem. Yet as the brainstem sucks, you cannot help but notice something like light entering and expanding within your body. The composition of this light is unknown to you, but we as readers are permitted to glimpse a little of its lecherific essence. For the first days, it isn't milk at all. It's a few tablespoons of colostrum, a golden mix of protein, immunoglobins, white blood cells, and laxatives that acts as a sort of inoculation and primer for the nascent digestive tract. Only around three days of age does the milk begin to come in, and it's a full two weeks before the transition is made to mature milk.

From a purely compositional perspective, this milk is spectacular stuff. It contains all the water, fat, sugar, protein, and minerals necessary to nourish you to the age of two without the supplementation of any additional food, and its specific formulation changes depending on your needs, so that if you have been born prematurely and need to put on weight, it will contain higher proportions of protein and fat. Even during a single feeding, the composition of the milk changes as foremilk becomes hindmilk, which, with two to three times the fat, cudgels you sweetly to sleep.

More startling than its nutritional composition are the immunological components of milk. Your mother's breast milk contains an ungodly quantity and assortment of immune factors, including cytokines to stimulate the thymus, chemokines to stimulate development of the intestine, glycoproteins like lactadherin that prevent rotaviral infections, mucins that block Salmonella and Norwalk virus, oligosaccharide prebiotics to help you select from your environment a beneficial microbiome, as well as decoys that bind to pathogens.

Some of the billion or so leukocytes that get passed to you in each feed even transform themselves into dendritic messengers for your own immune system. When these cells encounter a pathogen, they disassemble it and carry the pieces to your immature thymus, where they present the bits on their surface to instruct and alert the T-cells that will lead your own adaptive immune response. In this way, the milk not only protects you but also guides the development of your own immune system, and though it's important not to anthropomorphize blob-like single-cellular components, it's hard not to see these dendritic cells in a very parental role, offering to your T-cells what they've come to learn of the dangers of the outside world.

It is heady stuff, in other words, this original experience of nourishment. With the breast jammed against your face, your body releases serotonin and dopamine, as well as oxytocin (not to be confused with OxyContin, the prescription opioid), a chemical that causes feelings of wellness, of belonging, of attachment, and also sleepiness. Drinking milk is so completely satisfying that children commonly nurse not only to sleep but also while sleeping, meaning that milk is, quite literally, the stuff of dreams. It's a feeling of well-being so powerful that many infants can only settle when fed with a breast or bottle, or by mimicking the experience with thumbs and pacifiers, and when children must be weaned from the nipple or from its various placeholders, one form of comfort is often replaced by another. Some folks I know held a burial for their daughter's pacifier on her third birthday, giving the whole process a ceremonial flare and rewarding the girl's stoicism with a stuffed animal, in the hopes that, by substituting a new thing for a beloved one, they could prevent the kid from freaking out. And it is this transference of longing, according to some Freud-influenced theorists of the developing mind, that leaves some people feeling a perpetual emptiness in their lives, dooming them to forever sift the landscape for a suitable simulacrum, so that they trade the teddy bear for a dog, the dog for a guitar, the guitar for marijuana, marijuana for an MFA, the MFA for a spouse, all in a fruitless hunt for one single object or activity that can fill them again with the unspeakable contentment once experienced drinking milk. And who doesn't find, in the fraternity boxed wine game of Slap the Bag, some echo of the rhythmic smacks of the infant hand on the breast of the mother?

Though a lot of Freudian theorization is quackery, there is

something to the idea that we forever miss the bottle or breast. As the infant grows, it stops rooting and sucking in response to simple stimuli, but the brainstemmy reflexes that sustained its early survival don't exactly disappear. Rather, as the front of the brain develops, it learns to suppress them. The lobes that we use to calculate tips and remember birthdays and write ill-conceived poems to high school crushes, the part of the brain that we think of as containing *us*, who we are, this part gets progressively larger and stronger, and as it does, it extends its domain over the brainstem, quietly bringing more and more of these reflexive behaviors under its control. The front of the brain learns to silence the reflexes—but it cannot actually eliminate them. All throughout our adult lives, the brainstem continues to try to nurse. You may be calculating the mass of a boson or scrubbing the kitchen floor, and meanwhile, the brainstem is still dashing off its frantic mammalian diktats with the hope of sneaking past the censors in the frontal lobe.

This isn't simply some theory deduced from MRIs. You can actually see it happen, as it does sometimes in neurologists' offices and nursing homes and IC units across the country. As diseases like Alzheimer's and Parkinson's dismantle the more person-y areas of our intelligence in the front of our brains, they weaken frontal control over the brainstem. One of the ways of evaluating the extent of the damage is to brush your finger across the lips of a person. I have seen this happen. Perhaps you have too, on someone you love. As the finger brushes the lip, the mouth, unbidden, puckers, the neck cranes, and the human being you know falls away like a cloak, revealing, for a split second, a newborn mammal that has never, in some senses, stopped wanting its mother.

Nature's Mischief

In 2011, Alicia Tondreau-Leve left her husband, Alan, in Massachusetts and moved with her two sons to Florida. It's hard to say why. In a deposition taken years later, a witness for the State of Florida recalled hints of marital strife, which wouldn't have been unreasonable, given the couple's recent bankruptcy and foreclosure. But this hardly explains the decision. Every marriage has its quota of unpleasantness, but few people address it by placing 1,300 miles of I-95 between themselves and the life they once lived. I've spent a great deal of time talking with Alicia Tondreau-Leve about the way she reinvented herself in the Sunshine State, and in all our conversations, I was never able to get a satisfactory justification for the move. It was an opportunity to start fresh, she told me, while Alan tried to find a new position. Her father had retired to Treasure Island, Florida, she said, so she had some familiarity with the area, and it seemed natural to her to go where he was. These are reasonable considerations, but I can't say I entirely believe they justified such an extraordinary move. Treasure Island is on the Gulf of Mexico, after all, and she chose to live on the Atlantic side of the peninsula, hours away from her father or anyone she knew. I'm inclined to think the relocation to Florida and all the havoc that followed that decision had less to do with a troubled relationship or a retired father and more to do with the life she and Alan had built for themselves in Massachusetts and how it felt to lose that life. Foreclosure is a slow process, and they did not officially lose the home for another five years, but in all her trips back to the state during that time, Alicia Tondreau-Leve never slept under that roof again.

There's support for this idea, as well, in her approach to Florida. She didn't go, like so many would-be Floridians do, with some ill-conceived vision of sand and sun and lotion-slathered ease. She moved with her sons to Brevard County, settling in a rental home in one of the endless inland exurbs south of Cape Canaveral. No ocean views or even breezes, just an interminable skein of highways, dotted here and there with Walmarts and Walgreens and freshly bulldozed lots baking to dust in the sun. She chose a home in a wine-country-themed development that was for all intents and purposes a retirement community, but this didn't bother her. The schools were good. That was all that mattered. She planted herself on Chardonnay Drive, enrolled her boys, and got to work.

She couldn't work in a traditional job and still take care of her sons, but before she'd left Massachusetts, one of Alan's cousins had given her an idea for how to make ends meet: she could buy and sell baby formula, the powdered milk often relied upon by parents to feed their children. The cousin ran a consignment business in Massachusetts, and he had some experience with the market. All she had to do, he explained, was make a post on Craigslist saying she'd buy cans of unopened unexpired formula. Parents who were switching their children to solid food would then reach out to her. She could gather this formula, ship the cans to him, and he'd resell them through his store. He even gave her a template to use for Craigslist.

She decided to give it a shot. She put up the post online, and she printed flyers and started pinning them up at laundromats and children's consignment stores around Brevard County. Then, after dropping her boys at school, she started crisscrossing the county in her car, grabbing whatever she could lay her hands on. She enjoyed the freedom of it, being in charge of her own schedule, driving those wide Floridian highways, but it was hectic. At times, she'd be two hours from home, and she didn't know what she'd do if the school called to say one of her sons was sick and needed to be picked up. She had no friends or family in that part of Florida, no network of support at all. There wasn't much profit in it at first, either—at best a dollar or two a can. Factor in the time spent driving, the cost of gas, and the shipping expenses, and it was hardly much of a wage. But she kept at it. She thought of it as a hobby. She scoured Craigslist at night, looking for people selling formula, and she plotted her routes so that a longer drive would be offset by multiple pickups. She started making trips to Tampa, Orlando, Fort Myers. And little by little, word spread through the mom grapevine in Brevard County and beyond that there was a woman named Alicia who'd buy your leftover formula.

At this point, it wasn't clear when or even if Alan would be able to find work in Florida and join them. She knew she needed to increase the amount of business she was doing if she wanted to be able to support her family, and it wasn't hard to see what was affecting the bottom line. You couldn't expect to make much of a living when you were driving an hour each way to pick up five cans of formula and then spending ten dollars to ship them north. If she wanted to make the work profitable, she needed more formula. More units would mean more money, of course, but it would also bring down shipping costs per unit. But it wasn't like formula grew on trees. Though close to a quarter of a million babies are born in Florida each year, there were only so many people in Brevard County with extra cans. If she wanted more, she needed to expand her buying radius. She obviously couldn't drive all over the state herself, but she often met other parents in the course of buying formula. These were mostly mothers like herself, women with drive and ability who needed to make a living but had obligations at home that kept them from working the traditional nine-to-five.

Before long, she'd created a small network of buyers across much of the state. She had Giulyanna Guzman in Orlando, Angel Castellanos in Miami, April Engman in Fort Myers. In Brevard County, she had Sonya Barbour, and in Tampa, a woman named Alexis Dattadeen. She kitted them out with everything they'd need: branded business cards, the Craigslist template, price lists of what to pay for the various kinds of formula, tally sheets to keep track of exactly how much they'd gathered of each variety, boxes and bags for transporting the containers in the most space-efficient manner possible. She also gave them advice on when to buy formula—many parents tended to sell their formula at the beginning of the month—and on transacting safely with people met via the internet—meet somewhere public, somewhere with cameras. They were not direct employees, but independent contractors. They were, in essence, entrepreneurial miniatures of her, all doing exactly what she was doing, meeting up with other mothers, buying surplus formula. The only difference was that once they'd accumulated a sizable batch of product, instead of trying to sell it themselves, they'd sell it to her.

The operation grew rapidly. By 2012, her neighbors had begun to complain about the tractor trailers picking up deliveries at her home, so she rented a storage unit in Rockledge, Florida, just to hold the cans that were awaiting shipment. Soon it was two storage units. She outgrew Alan's cousin's consignment shop, but by then she'd found larger buyers who were able to offer better prices. She was drop-shipping formula to distributors in New York, Minnesota, sending whole pallets of the stuff to California by freight. She even developed enough reach within Florida that she was able to fill specific orders for her customers, a hundred cans of Gentlease, for instance, or twenty tubs of Good Start Soothe. There was so much to do that when Alan finally came down to join her, he started picking up deliveries, as well. By then, it was more than a hobby. It was a functioning business with branding and car decals and a steadily growing account at Bank of America. Around that time, she formally registered her company with the State of Florida, hanging up the paperwork at her storage unit. She called it Formula Mom.

The Floridian dream is a mirage for many, if not most. You think you're going to spend your days on the beach, but you end up living in a one-bedroom behind a swamp and working at a Verizon store in a strip mall. You only see the coast at night when an inebriated Brit at a beachside bar hails your Uber to take them back to their vacation rental. But Alicia Tondreau-Leve's dream had been different.

By 2014, she was sending out ever larger shipments. She'd connected with an exporter who was sending whole shipping containers of formula to China (a tainted formula scandal there had sent the demand for U.S. formula skyrocketing). All told, she'd sold more than a million and a half dollars' worth of formula. That summer she suffered a setback when her Tampa subcontractor, Alexis Dattadeen, decided to leave the business. But in the process of taking over Dattadeen's accounts, she connected with a supplier named Steve Riley, who was dealing in wholesale quantities. Things were looking up. One day, driving through an "old Irish"–themed residential community called Capron Ridge, she stumbled on a lot she liked on Tralee Bay Avenue and ended up sitting down with a salesperson. The ground had been leveled. The blueprints were already drawn up. He walked her through the design, making notes as they spoke. It was to be new construction—three bedrooms, vaulted ceilings, hurricane windows, a three-car garage, a lanai overlooking an honest-to-goodness stand of trees. She put down a $1,000 deposit to reserve the lot. The dream was in reach.

Alicia Tondreau-Leve still lives in Florida. She's fifty-seven now. She wears glasses, and her blond hair has been washed out with silver, but she still looks like she'd be comfortable sitting at a conference table in a pantsuit. She speaks in measured tones, parsing out each sentence until she's completed the paragraph in her mind. She can speak voluminously when a topic strikes her fancy, but she's circumspect on the subject of Formula Mom. When I emailed her to see if she'd be willing to chat about the company, her reply consisted of a single sentence, noting respectfully that she would respond after researching my credentials.

She is, in her own words, "a very reserved person," but she's also learned to be careful about exactly what she says and to whom. From 2012 to 2014, as Formula Mom was growing by leaps and bounds, she was the target of an extensive surveillance operation. Her Toyota Camry was tracked, her bank records subpoenaed. Her conversations and movements were covertly recorded on both video and audio, and one by one, people at every level of her organization were recruited as confidential informants. Steve Riley, the man who had promised to sell her formula in wholesale quantities, was in reality William Powell, an undercover officer in the employ of the State of Florida. So she has some good reason to be reserved about what she says and paranoid about who might be listening. She knows what it's like to have her words used against her, and moreover, there's the practical matter that whenever she talks on the phone or writes a letter or sends an email, someone really is monitoring her every communication. Because Alicia Tondreau-Leve is not living with her husband and sons in the three-bedroom with vaulted ceilings and hurricane windows on Tralee Bay. She's currently living with four hundred or so other women at 16415 Spring Hill Drive in Brooksville, Florida. She's serving out a twenty-year sentence at Hernando Correctional

Institute for running a criminal syndicate specializing in the theft of infant formula.

YOU MIGHT HAVE HEARD ABOUT THE FORMULA MOM trial a few years ago. It was a strange case, maybe even funny. People get busted every day of the week, of course, for buying and selling things they shouldn't. But when the detective's knife pierces the shrink wrap and the white powder that comes spilling out is not fentanyl but formula, that's the sort of center-cannot-hold tagline that's irresistible to local TV stations and national newspapers alike. There were a few splashy write-ups about Sonya Barbour and Glenn Martin, a couple in the lower echelons of the organization who had been caught stealing formula to support their pain pill addiction ("Florida Lovebirds Make $90,000 in Six Months Stealing Baby Formula"), along with a more sober-minded noirish follow-up in the *New York Times*. In the index to *The Decline and Fall of the American Empire*, which is being composed in real time in newspapers and on social media nationwide, Formula Mom became one of a hundred thousand entries under the subheading Florida.

In many ways, the blossoming of a registered and generally above-board business into a vast criminal enterprise seemed to represent the apotheosis of a distinctly American, distinctly Floridian brand of laissez-faire economics. And with a mild-mannered mother of two at the helm, cutting deals in between school drop-offs and music recitals, it also offered a seductive morality tale about the irresistible siren song of the United

States dollar. A lot of folks were very happy to take this story alongside twelve others from the daily doom-scroll and fit it into their own preexisting jigsaw puzzle of a United States falling apart at the seams. It was a story about the unraveling of the social fabric, the failure of traditional family roles, the debasing of the most fundamental human relationship. It was a story about moms not being moms. "For other frugal couples looking to save money on a newborn, there's always breast milk," wrote Mandy Oaklander for *Gawker*. "Though it doesn't sell quite as well."

This was, and still is, a complicated story. In a state devoted to no-holds-barred economic growth, a state reeling from the twin crises of a real estate bubble and a prescription drug epidemic, a foreclosed-upon mother enters a preexisting gray market for infant formula and uses the internet to create a network connecting sellers with buyers, and in so doing she turns the usual humdrum business of parents reselling their old cribs and strollers and formula for extra cash into a million-dollar crime ring. A lot needs to happen to get you from A to B in that situation, and there's a whole bunch of info the average reader, even if they drank nothing but formula for twelve months of their life, probably doesn't know a single thing about. What's a gray market, and why is there one for infant formula? Why is there an infant formula market at all? Why is there infant formula? There's more to the story than you might expect. Before you can begin to get a handle on the rapid rise and vertiginous fall of Formula Mom, you've got to get a handle on milk. If the story of Alicia Tondreau-Leve is a story about anything, it's a story about moms, and it begins 360 million years ago in a swamp.

IT'S A WET WORLD. THE WATERS OF THE PANTHALAS-sic Ocean span much of the planet's surface, an ocean so vast that one can travel around much of the globe without encountering land. The largest landmass, Gondwana, sits firmly over the southern pole, and a few continental-size peninsulas encircle the Paleo-Tethys Sea. The collision with Laurasia, which will eventually form the towering Appalachians, is only just beginning, and much of what will become North America is covered by endless marsh and shallow seas.

The world is mostly water, but the age of fishes has come to an end. Gone are the jawless galeaspids, with their bony headshields, charging across the shallow seafloor. Gone are the dart-shaped doryaspids and the ponderous, benthic psammosteids. In their stead, out of the lineage of the reptilomorphs, a new creature has taken her place: the amniote. Like the amphibian, she is able to leave the water, but unlike her cousin, she lays her eggs on land. To accomplish this feat, she's developed an adaptation that's never before been seen on the planet. Occasionally, to prevent her clutch from drying out, she crawls atop her nest and moistens her eggs with liquid exuded from pores on her belly.

This liquid is likely little more than water, but over time, natural selection will favor those mothers whose moistener contains inclusions—possibly agents that protect the egg from unwelcome bacteria, possibly proteins that penetrate the shell and nourish the developing embryo inside. And as the eons pass and the proteins in this liquid proliferate, the need will diminish for the yolk to provide nourishment. It will shrink and shrink, so the theory goes, before disappearing altogether. By this time, the lineage of the amniote will have expanded into a diverse group of life-forms, and certain environmental pressures will select mothers who incubate their young for longer and longer periods of time, until the egg disappears altogether, and they begin giving birth to live neonates, which they then suckle. And with this, an entirely novel class of animals will have come into being: animals that give birth to live young and use this nourishing fluid to continue the gestational period outside the mother. This class of animals, of course, will be named on the basis of the remarkable glands in the mother that provide this transitional nourishment, the mammals. And this is what makes it worth knowing about this little creature with her nest in the bog. She is not just any prehistoric creature. That liquid she dribbles on her eggs, that is the first milk. She is, in the mammalian sense of the word, the first mom.

 Now milk, let's just say this right off the bat, is a magnificent invention of nature. Fast-forward a few hundred million years from that swamp, and you have some sense of the evolutionary advantages conferred by being able to gestate babies on the outside of the body. Mammals have evolved to occupy ecological niches across the planet, and their milk has evolved along with them to suit their circumstances, whether it's the hundreds of pounds of milk that a blue whale mother feeds her calf each day or the rich milk of the Antarctic Weddell seal, so high in fat that at room temperature it's a solid. Not only has milk evolved but so has the manner of its dispersal, with manatees having nipples in their armpits and solenodon nipples flanking their anuses. There are short nursers—three or four days among the hooded seals—and long nursers—orangutans nurse sporadically for eight years—and prolific nursers—the two dozen nipples of the tailless tenrec come to mind. Some bats, in a terrifying spectacle of hooked nails and leathery wings, even nurse in flight.

Though milk is vital to all mammals, it's perhaps especially important to human beings. Considering that we give birth to some of the most helpless, most breast-dependent young in the entire mammal family, two different theories credit milk with our meteoric ascendance to the top of the planetary food chain. The first credits our outsize brains to this extended period of postpartum nourishment. There are limits, after all, to how long an animal can physically support the gestation of a fetus and how large a skull can pass through the birth canal. By prolonging that period of exterior gestation, according to this theory, humans were able to develop the larger and more complicated neural circuitry that is their hallmark. Piggybacking on this idea, another theory contends that this style of extended nursing and brain development endows humans with a kind of gestation that is fundamentally different from that of other animals. Because so much of the growth of the human brain takes place outside the womb, in the company of other humans, our brains evolved to be more socially attuned. This intuitive understanding of interpersonal dynamics, in turn, laid the groundwork for the creation of the complex social structures that are our species's hallmark. From milk, according to this concept, evolves society.

IN THE SUMMER OF 2012, ALEXIS DATTADEEN, ONE OF Tondreau-Leve's independent contractors, met with one of her repeat sellers in the parking lot of a Walmart in Palm Harbor to buy ninety tubs and eighteen cans of infant formula. Her seller brought with him someone Dattadeen had never met: a man named Donnie. In his forties, energetic and talkative, with a sort of biker persona, Donnie was not your usual parent-selling-formula type. In fact, his real name was George Moffett, and he was a Pinellas County sheriff's deputy who had vastly misjudged his undercover role. But bluster goes a long way. Donnie talked and talked, and if Dattadeen sensed something was awry, she did her best not to let it upset the transaction. Donnie quickly set to talking about how he'd gotten the $2,700 worth of formula that was about to change hands.

> **Donnie** *I mean, I've probably stole a lot of shit in my day, but this shit was hard, I tell you what.*
> **Dattadeen** *And you're the guy who's been getting all this all this time?*
> **Donnie** *I got a hookup that hooks me up.*
> **Dattadeen** *Do you have a number where I can call you?*

Dattadeen went through with the purchase, and over the course of that fall and winter, she met repeatedly with Donnie to buy formula that she believed to be stolen. Via text, she ordered specific formulas from him, prodding him to bring her as much as he could, and as the holidays approached, she asked whether he could also get her a pair of iPads for her children for Christmas. She committed countless crimes, but through it all, she evidenced not so much a hardened criminal mindset—her children were present during most transactions—as a galling innocence of the very criminality of her actions. In her conversations with Donnie, she spoon-fed him information on her business, telling him how often she met with buyers, how much she bought, what she paid, what sort of formula brought the highest return, how she put down the backseats of her minivan so she could pack it all in. Perhaps the best example of her criminal naivete: she once asked Donnie whether he could shoplift a trampoline for her.

Dattadeen was also extensively surveilled during that time, and this surveillance corroborated everything she'd confided in Donnie. Again and again, she was recorded meeting individuals in parking lots—7:00 p.m. at McDonald's, 10:00 a.m. at Walmart, 3:00 p.m. at a RaceTrac gas station—and exchanging cash for formula. As for where the formula was going, she'd mentioned to Donnie that she had a partner in Orlando. She met the woman every Monday in Lakeland to give her the formula she'd collected. That November, deputies followed her on one of these trips, tailing her to a Circle K gas station in Lakeland. There they recorded Dattadeen meeting with an older woman in a powder-blue Toyota Camry. So much formula had been transferred to the Camry during the transaction, they noted, that it caused the shocks of the car to visibly compress. The older woman was Alicia Tondreau-Leve.

CHILDREN HAVE BEEN DRINKING THE MILK OF OTHER mammals for thousands of years, often from vessels made specially for that purpose. In the graves of Bavarian children dead three thousand years, there are small anthropomorphic milk vessels. There are clay baby-feeders from Rome and ingenious teapot-shaped bottles with anticholic vents. There are even pewter "bubby pots," from England, coffeepot-like devices where a rag attached to the spout and suckled by the infant wicked milk up from the interior.

It wasn't until the mid-nineteenth century, however, that a trio of innovations paved the way for the creation and widespread use of formula. Processes for evaporating and drying milk allowed for the large-scale production of shelf-stable milk products, while the creation of the first India-rubber nipple simplified the means of feeding. With the addition of an increased understanding of the nutritional requirements of infants, the stage was set for the creation of the world's first manufactured infant food: a mixture of cow's milk, flour, potassium bicarbonate, and malt formulated by Justus von Liebig in the 1860s. Though Liebig's "Soup for Infants" was not necessarily the greatest commercial success ("Sensible parents," wrote one critic, "will be content to leave the recipe for some coming race who may prefer art to nature"), the era of formula had begun.

The next hundred years saw steady advances both in our

understanding of human milk and in the processes used to replicate it, and by the 1960s, infant formula had officially become a big business. In the United States, the market was dominated almost entirely by three mammoth companies—Abbot Laboratories (Similac), Mead Johnson (Enfamil), and American Home Products Corp. But while advances in manufacturing and nutrition science continued to shape this market in ways large and small, it was the implosion of Baltimore that fundamentally and irrevocably altered it.

The city was in freefall in the 1960s. Amid the drive to desegregate Baltimore's schools and the riots that followed the assassination of Dr. Martin Luther King Jr., the White population had begun to flee the city for the suburbs. Meanwhile, Bethlehem Steel, the area's behemoth employer, was beginning its long, slow, lumbering decline. What had once been the nation's sixth largest city was dropping down the charts. As with almost any crumbling of institutional infrastructure, the brunt of this collapse was borne most acutely by the city's children, and David Paige, a young pediatric resident at Johns Hopkins Hospital, found himself writing prescription after prescription to counter widespread malnutrition in the city's infants. It was unpleasant work. Malnourished infants would suffer from not only neurodevelopmental delays, stunted growth, and poor school performance, but they would also be at higher risk for an array of adult diseases, from diabetes to hypertension. "The individual burden is lifelong," he wrote, "the cost is enormous, and the loss to society is incalculable." Faced with this, he had a minor epiphany: instead of prescribing medicine to treat the health consequences of malnutrition, he should simply prescribe the food that would prevent malnutrition in the first place.

He set up shop at a clinic for newborns in the Cherry Hill neighborhood of Baltimore, providing iron-fortified formula to low-birthweight children, and after receiving a grant from the federal government, he launched a statewide program that gave Maryland families vouchers to purchase formula and nutritious food. After his research indicated the vast benefits of the program, it caught the attention of a trio of senators on the U.S. Select Committee on Hunger. On September 26, 1972, after a yearlong standoff with then-President Nixon, the Women, Infants, and Children program, WIC, was born as a two-year pilot.

From a public health perspective, the outcomes of the WIC program were, by all accounts, astounding and long-lasting. In the near term, there were the anticipated decreases in anemia and infant mortality, but in the longer term, children who relied on WIC showed improved vocabulary scores and an increased ability to remember numbers. They were less likely to suffer from mental health conditions, less likely to repeat a grade. The list went on and on. On the societal side, a study from the General Accounting Office later found that for every dollar spent, the program saved three dollars in future medical costs. It was and remains one of the best examples of smart government spending. As an article in the *Baltimore Sun* later noted, the program was "practically fraud-proof."

These days, a little under four million infants are born every year in the United States. Half of them get their formula from the federal government.

THE SECOND BRANCH OF THE INVESTIGATION BEGAN a little before dinnertime on February 11, 2013, when a newly engaged couple, Sonya Barbour and Glenn Martin, put their seven-month-old into a stroller and headed for the Target in Melbourne, Florida. They were both around thirty at the time. Barbour was trim, with a high forehead and strawberry blond hair. Martin was more thickly set, bearded. He'd been working as a roofer and occasionally helping pour concrete for swimming pools. In the store, Barbour pushed a shopping cart while Martin pushed the stroller. They made their way to the formula aisle. There, as Barbour blocked the view with the cart, Martin busied himself at the shelf. They had developed their technique through practice, but this trip was to end poorly. As they made for the exit, they were detained by Target's Loss Prevention team. Under the stroller, concealed beneath a rumpled blanket, lay $534.64 worth of formula.

That evening, in an interview at the Brevard County sheriff's office, Sonya Barbour stated that in the summer of 2012, not long after her child had been born, she'd seen an internet post by someone calling themselves Formula Mom, who was offering to buy formula for cash. Barbour responded, and soon she had made the acquaintance of a woman named Alicia Tondreau-Leve. She was supposed to be working as an independent contractor for Tondreau-Leve, buying surplus formula from around Brevard County and selling it to Tondreau-Leve, but she and her fiancé had elected to steal new formula instead of buying leftover cans from other parents. They sold the stolen merchandise to Tondreau-Leve, typically twenty to forty cans at a time for $9 each. The next morning, Barbour signed an agreement to work for the sheriff's office as a confidential informant.

On Thursday, March 7, 2013, Sonya Barbour reached out to Tondreau-Leve to say she had formula to sell. By this time, judges in Pinellas and Brevard counties had signed warrants to attach GPS trackers to the vehicles of Alexis Dattadeen and Alicia Tondreau-Leve, so the Brevard County sheriff's office had no trouble following Tondreau-Leve's movements as she traveled to meet Barbour at a Walgreens in Rockledge. Barbour had been equipped by the BCSO with an audio recording device, and detectives were stationed in the area, recording video of the interaction. It was not the usual conversation one overhears during transfers of stolen property. In fact, the two women spoke mainly of Barbour's personal difficulties, in particular her struggles to regain custody of her children and to find a clean safe space to raise them. (It would come out at trial that Tondreau-Leve had been quite invested in the well-being of Sonya Barbour's family. She had put the family up for a month in a hotel, helped Barbour to buy a car, given clothes and hair clippers to Glenn Martin for a job interview. As Barbour would say at trial of her confidential informant agreement: "I felt like a big piece of crap for doing that to somebody with such a big heart.") However, for all this talk, from a prosecutorial perspective, the most important moment came when Barbour said she was scared out of her wits. The person she'd bought the formula from had told her he'd stolen it from a store.

"Well, you shouldn't have taken it," said Tondreau-Leve in response. Then she paid Barbour for the haul and drove away.

The investigation continued throughout that year, turning up Formula Moms in Orlando and Fort Myers, establishing the extent of their networks and recording deliveries to Tondreau-Leve. By this time, with so many jurisdictions involved, the Florida Department of Law Enforcement had come on board to oversee the investigation. The structure of the organization had by then become clear. Across Florida, formula was being stolen, or boosted to use the technical term, by a small army of people. Many, if not most, of these boosters were newly minted members of the state's opioid crisis. (Barbour and Martin were spending an estimated $200 per day on their pain pill addiction, and when investigators looked into the network of Giulyanna Guzman, the Orlando Formula Mom, they were unable to find one supplier who wasn't addicted to opioids.) Lured by the promise of a quick payout, these people stole formula from chain stores like Walmart and Publix and dialed up whatever number they found on their local Craigslist. Their most reliable buyers, as often as not, belonged to someone in the Formula Mom organization. While perhaps not explicitly soliciting it, Tondreau-Leve and her independent contractors were hoovering up and paying for this stolen merchandise so efficiently that they'd become go-to fences for many addicts. Can by can and tub by tub, in the trunks of Toyota Camrys and the backs of

Dodge Caravans, this stolen formula all made its way from a shelf at Target or Winn-Dixie to the stacks in Tondreau-Leve's storage facility in Rockledge. Almost any time of day in Florida, a Formula Mom could be counted on to meet up and trade cash for powder, provided it was before bedtime.

The state attorney was planning to charge Tondreau-Leve under Florida's RICO (Racketeer Influenced and Corrupt Organizations) Act, though she was far from what most people might have imagined as a mafia don. She was a homebody with a Toyota Camry. Her idea of fun was to detour through in-process housing developments and look at the empty lots. Unlike most RICO cases, Tondreau-Leve had also left a substantial paper trail. She'd been running the business through Bank of America accounts, and she'd used wire transfers to make payments. She'd done much of her work over email and text. There were invoices, price lists, branded business cards. It was a strange case. There wasn't really any question that Alicia Tondreau-Leve operated an organization called Formula Mom that purchased and resold formula—the business was registered with the State of Florida—and there was no doubt that some of that formula, if not much of it, was stolen. What they lacked was evidence that Tondreau-Leve knew just what it was she had created. They needed to show her knowingly buying stolen formula.

WHEN YOU HAVE A BABY, EVEN IF YOU'RE BREAST-feeding, you usually need some formula on hand, and at some point, you're probably going to have an extra can or two lying around. It's not unusual to find yourself buying or selling a couple of tubs of Enfamil on Craigslist, and when you do find yourself there, as Alexis Dattadeen and Sonya Barbour and Alicia Tondreau-Leve did, it doesn't have the flavor of crime. It feels like the sort of thing parents have been doing for ages: passing on the stuff your kid can't use anymore to another kid who can. It's just infant formula, after all. In fact, it's easy to become a player on the market without ever realizing you might be engaging in anything nefarious. I can testify to that, because I've done it. When my first daughter was born, she needed some specialty stuff that was nearly impossible to find locally, so I did what any other harried parent would do. I bought it online. The possibility that I might be doing something illegal never crossed my mind.

But there's a good chance that I was, and it has to do with WIC. Because there was one outcome that went unforeseen when David Paige first began to give away formula in Baltimore: the economic consequences of completely rejiggering a highly concentrated market. By taking many of the formula market's most price-conscious consumers and replacing them with a very obviously blank check, WIC profoundly shifted the sales landscape in the United States. The country's three main manufacturers of formula were quick to get the lay of this new and fertile territory, and over the following two decades, the price of formula steadily shot up even as the price of its basic ingredients—milk, vitamins—remained relatively static. It was a firehose of cash, the flow guaranteed almost regardless of prices. As Senator Metzenbaum of Ohio noted in a 1990 hearing of the Senate Subcommittee on Antitrust, Monopolies and Business Rights, the wholesale price of infant formula had increased 150 percent during the 1980s. The price of milk, formula's primary ingredient, had increased by only 36 percent.

Following a federal action against the country's three largest formula makers in the early 1990s, states across the country instituted a system wherein formula manufacturers would have to bid for the right to supply WIC formula within the state, and this did bring down costs for the government—the WIC program currently pays as little as 2 percent of the wholesale cost of a can. But the artificially high retail price of formula has remained.

When something costs more in one place than another, a gray market arises. Though the definition of a gray market, like the market itself, is a little cloudy, a gray market is one that facilitates any quasi-legal movement of goods through unauthorized distribution channels. Gray markets often crop up when there's a price imbalance between two countries. For instance, if a vacuum costs ten times as much in Argentina as it does in the United States, there's an incentive to get U.S. vacuums onto the Argentine market by whatever means available, and that goes a long way toward explaining what was happening with infant formula in the United States when Formula Mom started doing business. Inexplicably high formula prices, combined with inelastic demand, created a product with a reliably strong resale value. Add to this the slim margins made by independent retailers, and you had a buyer for the supply. Give those buyers and sellers an easy and anonymous means to connect—the internet—and all the pieces were in place for a thriving gray market.

This was the market where Formula Mom did business,

and there was really nothing wrong with buying leftover formula and reselling it. At some point, however—and it's not clear when this occurred or the degree to which Tondreau-Leve realized it was occurring—the business model shifted.

At any one time, after all, there are only so many parents weaning their children off formula, even in a state like Florida where hundreds of thousands of children are weaned each year. Assuming each of those parents has at most a handful of cans, to collect a meaningful quantity of formula, you'd need to connect with and arrange pickups from dozens of new sellers every single day. Even if those sellers were concentrated in densely populated urban areas like Tampa or Orlando, you could still easily spend ten hours just going from one meetup to the next. And you might drive an hour only to discover the formula you were promised was expired or two of the cans had a dent. The logistics of such a model are daunting. From the perspective of establishing a consistent supply, it's far better to work with the same sellers, people who know the routine and can regularly sell you set amounts.

Now, it's very much illegal to sell formula you receive through the WIC program, and Tondreau-Leve did include a note in her Craigslist posts that she would not buy WIC formula. But there's little doubt that quite early on, defrauding the federal government became central to the Formula Mom business model. Sonya Barbour and Alexis Dattadeen had both connected with Tondreau-Leve initially to sell her formula they'd received from the government. And let's be clear about this. No matter your standards of morality, defrauding a program devoted to infant nutrition is a pretty lowdown thing to do. But if you're a struggling mom, it's also not that crazy. WIC keeps sending boxes of formula, and your kid's not using it all. It doesn't feel like a deadly sin to sell a few cans for twenty bucks. It doesn't feel much worse to buy them for twenty and resell them for thirty.

THE SETUP WASN'T COMPLICATED. AS WITH ANY UN-dercover operation, from your everyday drug sting to the complex machinations of a Mr. Big investigation, the key was to gain Tondreau-Leve's confidence through a series of ever-larger sales while attempting to elicit some acknowledgment that the goods in question were stolen. The investigators were hoping to accomplish two things. First, should Tondreau-Leve attempt to argue that she was entrapped, they needed to show that she was "predisposed" to buy stolen formula. Second, and perhaps more important, they needed to show that Tondreau-Leve knew, or at least should have known, that she was buying stolen formula. If they could show that she deliberately avoided learning that the formula was stolen, they would meet the criteria for willful blindness, allowing them to make the case that she was dealing in stolen property.

On April 17, 2014, law enforcement approached Alexis Dattadeen with the evidence they'd gathered on her criminal activities, and in almost no time, she'd agreed to cooperate with the investigation as a confidential informant. They arranged for her to make a formula exchange with Tondreau-Leve while wearing a microphone and camera, and they gave her a script that they hoped would lure Tondreau-Leve into making a deal. Dattadeen was to tell her that she was no longer comfortable working for Formula Mom. It was too much stress, and she was worried about one of her new suppliers.

> **Dattadeen** *I'll find a regular job, a normal job. I mean, I'm out there every, you know, every day and I risk my—myself every time I go out there.*
> **Tondreau-Leve** *Why do you risk yourself?*
> **Dattadeen** *All these tubs.*
> **Tondreau-Leve** *What's wrong with them? Are they stolen?*
> **Dattadeen** *I don't know.*

They'd come from someone named Steve Riley, Dattadeen said, and he'd told her he could get an entire pallet.

The next day, Tondreau-Leve texted Dattadeen asking for Steve's contact information so she could deal with him directly. Dattadeen replied with the number for William Powell, the undercover operative with Florida Department of Law Enforcement. A week later, Tondreau-Leve met Steve Riley/William Powell in the parking lot of a CVS in Cocoa, Florida, to buy 67 cans of Similac for $790. They met again a week after that at the same CVS for a sale of 283 cans ($2,601). About two weeks after that, on June 5, they met at a Citgo to exchange 150 cans for $1,752. At that meeting, Steve told her he would be getting 3,000 tubs soon.

The final scene took place in a parking lot at a Lowe's Home

Improvement store. It was July 2, 2014. Two rental trucks pulled into the lot. Powell, who would later resign from the FDLE for his involvement in an illegal search, was driving one truck. In the back were six pallets of formula containing, all told, 3,300 tubs of Similac Advanced. The other truck was piloted by Alicia Tondreau-Leve. Her husband, Alan, arrived separately, in his white Ford Focus, carrying a white envelope containing $33,000 in cash. The trucks were backed up so their rear bumpers faced each other, and for well over an hour, on that scorching July day, as Alicia Tondreau-Leve and Alan Leve and another undercover agent masquerading as Steve Riley's wife all looked on, Powell moved the pallets with a pallet jack from one vehicle to the other. When he had finished, Alan handed the envelope to Alicia, who handed it to Powell, who handed it to his supposed wife to count. They were still standing in the back of Powell's truck when they heard the sirens.

THE CASE WASN'T HARD TO MAKE. ALONG WITH HAV-ing representatives from various supermarket chains itemize their losses before the court, the prosecution had rounded up an assortment of boosters willing to testify about their methods of theft in exchange for leniency in their own sentencing. Many of Tondreau-Leve's subcontractors—Dattadeen, Barbour, Guzman—were there to testify on behalf of the state. But a conviction still hinged largely on Tondreau-Leve's purchases from the undercover operative.

While the fact remained that Tondreau-Leve had bought 3,300 tubs of formula in that final sting, she had done so in broad daylight and had even brought with her a receipt for the delivery. Neither were her conversations with Powell by any means a nail in her coffin. Unlike Dattadeen, who had efficiently wrapped herself in a prosecutorial net of her own construction, Tondreau-Leve was harder to pin down, as her first text exchange with Powell made clear:

> **Tondreau-Leve** *I would like your business and would like to be your number one source you sell to. That being said, please confirm these cans are acquired legally.*
> **Powell** *Ha ha, is my word good enough?*
> **Tondreau-Leve** *Yes, of course, but I don't find my question humorous.*

And in her first meeting with Powell, having found out that the formula she was buying had come from Powell/Steve's brother, she asked where the brother got it:

> **Tondreau-Leve** *You made kind of like, you know, a ha ha, when I said, you know, "Are they legal?"*
> **Powell** *Oh, of course, you know.*
> **Tondreau-Leve** *There's a lot of people who steal them, so I just gotta be, you don't know that?*
> **Powell** *Well, I'm not really huge into that.*

And in her second purchase:

Tondreau-Leve *You're guaranteeing me that they're not stolen.*
Powell *Absolutely, Alicia. I guarantee you the world.*

In what would be one of her final deals, after becoming increasingly frustrated with Steve's hemming and hawing about the formula's origins, there was finally this exchange:

Powell *Well, I'll tell you whatever you want to hear.*
Tondreau-Leve *I want to know if it's stolen.*
Powell *No.*

This was murkier territory than most prosecutors would prefer to wade into. It was impossible to say whether this was a person trying to give herself legal cover while knowingly buying stolen merchandise or if she was instead doing her best not to cast aspersions on a reliable supplier. "Please don't be insulted that I check out who I buy from," she later texted Powell. "I have worked very hard for my business and need to protect it. I hope you understand." It was possible she was a crook. It was also possible she was just hearing, as Powell had put it, whatever she wanted to hear.

The final moments of the Formula Mom trial, even if you are only reading them in the court reporter's transcript, remain acutely unpleasant. It's early August 2016 in Polk County, Florida, a little after noon. The foreperson hands the verdict forms to the bailiff, who hands them to the judge, who looks them over and hands them to the court clerk. And the clerk reads.

"In the Circuit Court of the Tenth Judicial Circuit, in and for Polk County, Florida, case number 2014-CF-5929, we the jury find as follows . . ."

Though his role has been fairly minor, Alan Leve has been dragged into the case alongside his wife, and between the two of them, there are a number of counts, some with dozens and dozens of subcounts known as predicates. It's a long list to get through, but from the beginning, it's very clear how the jury has found. The mishmash of legalese is broken only by a steady punctuation of "guilty, guilty, guilty" and by Alan's exclamations.

"How could you do this?" he says. "Oh, God."

The clerk continues, ". . . Predicate 41, Predicate 42, Predicate 43, Predicate 45—"

"It's okay," says Alicia.

"—Predicate 49," says the clerk. "Predicate 50A, Predicate 50B, Predicate 51A, Predicate 51B, Predicate 52—"

"Oh, God," says Alan.

Alicia maintains her composure a few minutes longer, but when they find her guilty of conspiracy to commit racketeering, she falters.

"Oh, my God," she says.

YOU CAN'T SPEND MUCH TIME WITH THIS CASE WITHout beginning to get a strange feeling. On the one side were Giulyanna, Sonya, Jennifer, Janine, April, Alexis, Alicia. On the other were Kevin, Todd, George, William, Paul, Nicholas, and the Honorable Judge of the Tenth Circuit, Wayne. There were exceptions, of course. Alan helped with the business, and a man named Angel was one of Alicia's top subcontractors. And the state attorney general was a woman named Pam Bondi, better known at that time for rescheduling an execution so she could attend a fundraiser and for declining to pursue a fraud investigation of Trump University after receiving a $25,000 donation from the Trump Foundation. But by and large, if you were to take away the names and simply refer to every person by their gender, this was a story about a bunch of women being investigated and prosecuted by a bunch of men. It's the sort of stark delineation by type that gives any person an uneasy feeling that the justice system, in its rush to use on moms an apparatus designed for drug lords, had missed what this was all really about. As Powell, a highly trained undercover operative, testified at the trial, he took part in the sting operation without actually knowing much about formula.

"In the real world?" the prosecutor asked. "Like as a person?"

"Here and in the real world," said Powell.

That's an important gap, obviously. You don't go undercover on a heroin bust if you don't know anything about heroin. But it's more important because formula isn't just some lesser substitute for human milk. Yes, it feeds infants, but that's only part of the reason it exists. It also exists because, despite all the "Breast is best" slogans, breastfeeding for any extended length of time is an economic disaster for the person doing it. Even if you're not working as a bus driver or an Amazon picker, even if, thanks to the Affordable Care Act, your workplace has mandatory lactation facilities, breastfeeding for the CDC's recommended twelve months still causes a mother's lifetime earnings to plummet. Formula exists because it's a mom's best shot at equal footing with her male counterparts.

So, what you're really saying when you admit to not knowing much about formula is that you don't know much about what it's like to be a mom, and that's a big deal, because the real question here is what exactly in our culture turned a bunch of largely law-abiding mothers into felons. And it wasn't anything that complicated. These were women who'd been sidelined from traditional employment by their parental obligations. They were trying to make ends meet while raising kids. Formula just happened to be near at hand.

THE WORD MAMMAL COMES TO US FROM CARL LIN-naeus, the Swedish naturalist best known for using binomial nomenclature to organize the jumble of species on this planet into a cohesive system. It was during a six-month stint in the far north of Scandinavia, a time when he subsisted on little more than reindeer milk, that the idea for this taxonomy first struck him. If he knew how many teats each animal had, he wrote in his journal, and where they were situated, he would perhaps be able to contrive a natural system for the arrangement of quadrupeds. Years later, refining his *Systema Naturæ,* in search of a word to describe that class of warm-blooded vertebrates who give birth to live young and feed them outside the body, he adopted the Latin term *mamma,* meaning "breast." In this sense, for all its scientific veneer, to be called a mammal is really not much different from being called a breastie or a boober.

Though that doesn't totally capture what it means to call yourself a mammal. The *OED* hypothesizes the Latin *mamma* itself is only a mutation of a word that predates writing: *mama.* In which case, to consider yourself a mammal is really only to call yourself one who has been mothered.

As for where *mama* comes from, the composer Leonard Bernstein had a hypothesis. In his 1973 Norton Lecture at Harvard, he described once staying up all night trying to figure out the original language of human beings:

> I tried to imagine myself a hominid and tried to feel what a very, *very* ancient ancestor of mine might have felt and might have been impelled to express verbally. . . . I began by imagining myself a hominid infant just lying there contentedly trying out my newfound voice—*mmmmmm*—and then I got hungry—*Mmmmm! MMMMMMM!!!*—calling my mother's attention to my hunger, and as I opened my mouth to receive the nipple—*MMMAAAA!*—lo, I had invented a primal word: Ma.

He's wrong, of course. Babies don't go *MMMMMMM!!* when they're hungry. They can't because their mouths are open and they're screaming. Anybody who's held a hungry baby for two seconds knows that. But he was sort of on point about *ma.*

Ma does tend to be one of the earliest infant vocalizations, and some linguists believe it develops from the sounds made while breastfeeding. One only needs some syllabic reduplication to produce *mama*.

It's no coincidence then that *mama* is the first word of many children. It is a word that forms itself, called forth at the physical conjunction of a child's hunger with the thing that eases it. In which case, *mama* and *mamma* are largely one and the same. The desire for nourishment, through some archaic infantile metonymy, becomes a term for *milk*, for the organ that produces it, and for the person who bears that organ, and Camille Paglia is maybe more right than she knows in saying women are "nothing more than milk sacs."

Despite his misunderstanding of the sounds of a crying infant, Bernstein's right in a way too, in that *mama*, thanks to its uncanny recurrence in languages across the globe, is now hypothesized to be one of the earliest examples of human speech, perhaps even the first word. Which blows the mind a little bit if you think about it. It is our ability to communicate with subtlety and nuance that makes us special on this planet. To think that language, the very foundation of human society, might all be an outgrowth of some long-ago infant yearning to be fed gives the act of speaking a weird and slightly forlorn air. Every word is shaped a little like a nipple.

It's hard to say then, considering the word's presumed place at the very genesis of language, what exactly it means to call yourself a mom, let alone a Formula Mom.

Alicia Tondreau-Leve is still on Spring Hill Drive, folding

laundry and awaiting news of her most recent appeal. There's plenty that could weigh on her now. She owes close to three-quarters of a million dollars to the State of Florida, $500K in fines for her offenses, and another $250K or so for the costs of investigating and prosecuting them. For a long time, to satisfy this outstanding debt, the state confiscated any funds sent to her by her family, making it difficult to get access to necessities like toothpaste and soap. Her husband, Alan, was released early from Marion Correctional Institution in December 2020 after being diagnosed with late-stage bladder cancer. He died less than four months later.

But what Alicia Tondreau-Leve thinks about most anymore are her sons. That August evening, after the trial ended, the boys, by that time in high school, were in some senses orphaned. The younger of the two went to live with his music teacher, the older stayed off and on with his grandfather, the retired police captain, on Treasure Island. Their lives went on. They are no longer children—they're in their twenties now. They're good boys. One's a musician, the other works in finance. They visit when they can, and when they can't, they talk to their mother on the phone. No matter when she calls, whether they're in class, in a meeting, at dinner, they pick up. No matter what, they pick up. The calls, facilitated by ViaPath, an outside technology vendor, last exactly thirty minutes and sound like they are being transmitted through a long felt tunnel. The boys talk to her about their lives and about her appeal, and at the end, after the disembodied voice of a young woman cuts in to announce there is one minute remaining, they say goodbye. "Bye," they say. "Bye, Mom."

VII

Flatulence

If brass wakes up a trumpet, it is not its fault.

Arthur Rimbaud

Terms

n. a toot, a poot, a ripper, a fart, a fartsel (a small flatus, see turdlet), an SBD (silent-but-deadly; alternately SBV, silent-but-violent), an LBH (loud-but-harmless), a bunny, an airplane, a duck, a stinker, gas

Segment of the thirty-foot-long scroll
He-Gassen, or "Fart Battle," 1846

Biological Prologue: Flatulent Signatures

Flatulence is difficult to study. First and foremost, it's hard to get at. While capturing breath from the mouth and nose is fairly easily accomplished, the anus, socketed away between the buttocks and the perineum like a still in a brambly Appalachian holler, requires a great deal of Mylar and ingenuity. The second difficulty: gas escapes. While a glass jar may seem solid, given time, the low–molar mass atoms in a flatus—hydrogen, for instance—will pass right through it. Third, the compounds responsible for flavor and scent are susceptible to all sorts of chemical changes, from oxidation to thermal degradation, reactive processes that are only accelerated when the molecules are in a gaseous state. It's not just hard to get a flatus in a jar; in other words, it's impossible to keep it there. Despite these challenges, some research has been accomplished, and though much of it yearns for peer review, we can begin to make out the contours of our flatulence.

Maybe the most fundamental question in flatulence is who authored a particular flatus. Considering that the human nose has about four hundred smell receptors, each of which can be activated in combination with others, endowing the nose with the ability to identify a repertoire of odors vast enough—one trillion is the theoretical limit—that its extent remains a matter of debate, one would expect that any given flatus could be linked with near certainty to a particular person based entirely on an investigation of its smell. And to a degree, in a nonscientific sort of way, it can be. When I was in middle school, for instance, it was easy to identify the flatulence of a certain friend by the way his fairly innocuous cabbage-based flatus, chock full of low-molar gases, effused rapidly through his jeans, in the process liberating and carrying with it the distinctive scent of the All detergent his family used. Generally, though, humans are poor at recognizing scents.

There's a temptation to turn to the artificial intelligence technology that allows for facial and voice recognition, but while AI has in many ways mastered sight and sound, it has no corollary in the world of smell. Though artificial neural networks have learned to identify smells, they can only do so using data that represent odors, not the odors themselves. For computers, the actual smell of flatulence remains, so to speak, in the jar.

There's promise in adapting voice recognition technology to the identification of flatuses, but here, as recent research has shown, there's an unexpected difficulty: while different flatuses might sound remarkably distinctive to the lay ear—from a deep post-Thanksgiving blubbering to the drawn-out squeal of a balloon with its neck stretched taut—they are all in fact remarkably similar. Adult human flatulence tends to have its maximum power in a short burst in a narrow range between 200 and 300 Hz, with a distinct peak right around 275 Hz, very near C# in the fourth octave (think of the final note of the descending motif in Bach's Prelude and Fugue in C-sharp minor). The perception that certain flatulence sounds higher pitched is due to the biases of human hearing,* which is less sensitive at lower frequencies so that high-pitched sounds are perceived as louder (the memorable C# to C# octave leap in the Prelude gives some sense of how much louder higher frequencies can sound).

Despite this, however, researchers have achieved some progress. A database of thousands of audio recordings of episodes of flatulence, warped using the mel-frequency cepstrum to better mimic the response of the human auditory system, was used to train a neural network with a three-layer architecture (fifty hidden units, weights optimized via a scaled conjugate gradient backpropagation algorithm). Not only did the network prove adept at distinguishing flatulence from other sounds, but it was even able to pair flatus to deflator with a high degree of accuracy.

So successful was the work at identifying flatulence that in 2021 an algorithm was unleashed upon the internet, scanning publicly available audio in an effort to detect, by its 275 Hz Gaussian signature, the unrecognized flatulence of the human race's recorded history. Its work has not been without reward. Within months, the algorithm detected a near-certain flatus in

---

* Combining research on the sound of flatulence with information about the concentration of VOCs (volatile organic compounds) in a flatus also uncovered a secondary but no less important understanding of the relationship between perceived loudness and perceived odor. While louder flatuses tend to contain more VOCs, a fact that seems to controvert popular wisdom regarding what are often called silent-but-deadlies, VOC levels were also inversely correlated to frequency, with a lower-pitched flatus producing more VOCs than its higher-pitched equivalent. This means that for any two flatuses of the same loudness, the smellier flatus will be perceived as quieter, a phenomenon most notable at the lowest end of the auditory spectrum, where a truly VOC-laden flatus might indeed slip below the auditory threshold.

an interview from 1950. "All should be free to make the best of themselves, and their country," says Winston Churchill in a brief bootstrappy monologue. "To rise by their own exertions in the life of a free and active society." It is the word *exertions*, pronounced, by Churchill at least, with an abdominally intensive near-onomatopoeic grunting, that provokes a faint but still definitive blip in the data, a just barely audible energetic burst of around 275 Hz.

Rites of Passage

Notes and remarks concerning the customs and practices of the Preadolescent Male in the Commonwealth of Pennsylvania, condensed and analyzed, with relevant commentary

For Captain Sir Richard Francis Burton
KCMG FRGS RIP

1. Background of the Gastric Process

And then he, with immense force, produces the happy farts, which I most humbly think may, with great delicacy and propriety, be fairly called a sixth species of farts.

Charles James Fox

Flatulence in the historical record is inescapable. Members of the Academy will recall not only the flatulence of Amasis that unsettled the twenty-sixth dynasty in Egypt but also the seder-flatus of the Roman soldier in Jerusalem that fomented the Jewish-Roman Wars, resulting in the destruction of the Second Temple and the scattering of the Jewish people. We are all familiar, as well, with the *He-Gassen* of medieval Japan, prints of which hang above the dais in the East Library. Likewise, as our colleagues in theology report, it is clear that the Protestant Reformation could hardly have taken place without Luther's determined belief that the devil fled flatulence, so that flatus itself became an emblem of antipapal sentiment specifically and theological revolution generally, as we see so often in the Academy's fine collection of woodblock prints by Cranach the Elder. Perhaps it is this Protestant influence, in conjunction with the pragmatist flatulence of Franklin[*] and the atmospheric worldview of Priestley, that gives flatus its outsize role in the culture of the Pennsylvanian.

But preliminary to any extensive investigation of flatulence qua social custom, it's necessary to provide a résumé of some of the salient scientific literature relating to intestinal function, despite or perhaps because of the familiarity of Members of the Academy with this subject matter, and in so doing to correct a number of erroneous conjectures and common misconstruals on the topic that might otherwise prevent a complete understanding of the present lecture.

---

[*] "What Comfort can the Vortices of Descartes give to a Man who has Whirlwinds in his Bowels! The Knowledge of Newton's mutual *Attraction* of the Particles of Matter, can it afford Ease to him who is rack'd by their mutual *Repulsion*, and the cruel Distensions it occasions?"

EARTHLY MATERIALS

Considering its role in ensuring physical and mental health, and through these the well-being of society as a whole, the function and dysfunction of the gastrointestinal tract is, as our biologists have long averred, vastly understudied,[*] and yet enough of the basic data has been elucidated to give a cursory explanation of the physical action of expelling gas. The process begins, of course, with the ingestion of food. The quantity[†] and quality of the flatus is the result of interactions among the genetics of the particular person, the genetics of their microbial community, and the type of comestible eaten.[‡] Certain foods—dairy, whole grain breads,[§] members of the brassica family, beans—contain molecular structures that our diges-

---

[*] See Part III of this treatise, On the Dearth of Scientific Literature in the Enteric Oeuvre. Countless questions remain outstanding. What precise peristaltic function controls the passage of intestinal gas, and is it equally functional when the subject is standing, supine, in microgravity? How precisely does the medulla oblongata process afferent data from the anal sphincter relating to physical sensations such as "wet," "heavy," and "spicy"? What is the coevolutionary history of the enteric and central nervous systems, and what does the networking of these two systems reveal about human intelligence as an evolutionary quantum?

[†] Roughly 30 to 400 mL per flatus, 500 to 1500 mL per day, with the largest specimens occurring upon waking, indicating a Total Daily Pennsylvanian Volume (TDPV) of roughly thirteen million liters.[1]

[1] In his essay *On Dreams,* Sir Thomas Browne mentions that certain foods have the ability to give "turbulent" dreams, noting that Pythagoras "totally abstained from beans" while Cato "doated upon cabbage" and "Daniel, the great interpreter of dreams," favored a "leguminous diet." The dreaming mind, in rapture to these indigestibles, enlarges ordinary nighttime events to extraordinary proportions: "a small puff of wind [makes] a tempest ... a spark in the bowels of Olympias a lightning over all the chamber."

[‡] The diet of the Pennsylvanian is somewhat merciless in this respect, encompassing the cabbage-swaddled Golabki of Pittsburgh, the polysaccharide-riddled Cheez Whiz steak sandwiches of Philadelphia, and the lima beans, the hoagies, the scrapple, and trout of central Pennsylvania.

[§] Pumpernickel, a favorite among some descendants of the Pennsylvania Dutch, carries the conjectural etymology of "devil's fart," though the Oxford English Dictionary is silent on this account.

tive systems are incapable of breaking down on their own, so the work of digestion falls to microbes within the gut. These bacteria devour pectins, raffinose, hemicellulose, and lactose, among other substances, symbiotically producing various nutrients and metabolites for their hosts, while generating a gaseous mélange, primarily (>99%) made up of the odorless gases oxygen, nitrogen, carbon dioxide, hydrogen, and methane (the latter two being notably flammable[*]) and including trace amounts of volatile sulfur compounds (VSCs) such as hydrogen sulfide (H_2S), methanethiol (CH_3SH), and dimethyl sulfide (CH_3SCH). These gases give the flatus its distinctive odor,[†] with the corrosive H_2S providing the scent of "rotten eggs," CH_3SH imparting a "rotten vegetables" aroma, and CH_3SCH striking the dissonant "sweet" note. Though the flatus of women has higher concentrations of these volatile compounds, the larger volumes of male flatus ensure average VSC emissions are equivalent.

---

[*] One female elder recalled a young man from her youth who would too often entertain gathered friends and family by leaning back in his seat, raising one corduroyed leg into the air, and igniting jets of his internal gas in extravagant bursts with a Zippo lighter kept on his person expressly for the purpose.

[†] While experimentation has confirmed the algorithmic possibility of identifying subjects based upon the sound of their flatulence, no rigorous attempts have been made to identify subjects based upon the odor of the flatus. Anecdotal evidence suggests it may be plausible, however. (Members of the Academy will recall the screening of Mike Myers's *Goldmember*: "Oh, everyone likes their own brand, don't they? This is magic....")

The determinants of flatal sillage[*] remain entirely conjectural.

The accumulation and locomotion of this gas in vivo is overseen by the enteric nervous system, two tubes of neurons, one nested inside the other, stretching from the esophagus to the anus. These tubes interact with the brain via the lowest part of the brainstem, the medulla oblongata, and also through the vagus nerve, sending information about the bowel's contents and receiving conscious and subconscious data in return.[†] But this fiefdom of five hundred million neurons has enough neuronal firepower and ganglionic independence that it can function even when the vagus nerve is severed, earning it the moniker of "the second brain."[‡] It is the enteric nervous system that registers the distension in the small intestine and relays those feelings of gastrointestinal discomfort to the brain, where they are processed semionomatopoeically, in the con-

---

[*] The capture and/or dissemination of flatus is itself a fascinating topic, covered in further detail in Part VIII of the current treatise. Of note are the charcoal-studded gas-tight Mylar pantaloons fabricated by Suarez et al. to capture pinto bean–induced flatus, as well as the bathtub, beaker, and syringe collection system of Tangerman.

"The volunteers submerged the lower part of their body in a warm bath each time they felt an urge to deflate. The flatus emission was completely sampled by catching all the flatus bubbles in a measuring beaker, which was submerged, fully filled with water, and turned upside down before the emission. After some practice, it is very easy to quantitatively sample all the bubbles."

In addition, Tangerman astutely observes the dilution factor inherent to flatus as well as three of the five flatus thresholds: the perception threshold, the 100 percent recognition threshold, and the threshold of objectionability. He omits the thresholds of utter repugnance and unbearability.

[†] So that, upon glimpsing a cheesesteak, one's stomach begins to grumble.

[‡] In all reality, the enteric nervous system's five hundred million neurons hardly justify comparison to the hundred-billion-strong goliath that is the human brain, but it's worth noting that the gut's half a billion neurons give it roughly twice the operational capacity of the average house cat's brain.

genial tongue of the Pennsylvanian, in terms of degrees of "bloat."*†

It is also the enteric nervous system, through actions as yet undescribed by science, that massages gas down the length of the gastrointestinal tract. As the gas, steadily or sporadically, fills the rectum, it exerts more and more pressure on the anal sphincter, which in turn alerts the brain. Should the psychological discomfort‡ of releasing this accumulated gas outweigh the physical discomfort of retention, the rectum may hold its contents for an extended period of time,§ ¶ but eventually it must

---

* As in "I overdid it on the mozzarella sticks and now I'm all bloated."
† The sensation of bloating is not, as is commonly assumed, due to larger-than-average volumes of intestinal gas.[1] Rather it is caused by gas pooling and applying heightened pressures in certain parts of the GI tract along with heightened afferent sensitivity in the mesenteric plexus, and it shows correlations with various disorders of the bowels.

[1] However, it is true that the lower pressures of high altitude cause increased air volumes in the bowels, and likewise is it true that because of the heart arrythmias of Apollo 15, the astronauts on Apollo 16 suffered increased flatus volumes due to the inclusion of potassium supplements in their orange juice:
128:50:37 Young: I have the farts, again. I got them again, Charlie. I don't know what the hell gives them to me. Certainly not . . . I think it's acid stomach. I really do.
128:50:44 Duke: It probably is.
128:50:45 Young: (laughing) I mean, I haven't eaten this much citrus fruit in twenty years! And I'll tell you one thing, in another twelve fucking days, I ain't never eating any more. And if they offer to sup(plement) me potassium with my breakfast, I'm going to throw up! (pause) I like an occasional orange. Really do. (laughs) But I'll be durned if I'm going to be buried in oranges.
‡ Those suffering from Olfactory Reference System believe their person is acutely malodorous, but it is not clear whether this is merely a context-specific symptom of a larger underlying psychiatric illness or its own classifiable disorder.
§ *Guinness* here is silent.
¶ It is scientifically accurate but not spiritually frank that a fart, if held in long enough, may eventually escape the body as breath; hydrogen is reabsorbed into the bloodstream in the intestines and colon and returned to the lungs, where it can then be exhaled (not burped), but the sulfurous essence of the fart makes no such migration.[1]

[1] Oxygen too is absorbed by the colon, a potentially lifesaving therapy now being tested upon asphyxiated mice.

loosen sufficiently that air, though no matter, escapes. At this moment, traveling at no more than .098 m/s,* a flatus is born.

2. Flatulent Milieu of the Pennsylvanians, Including the Modeling Effects of Breeding Adults and Elders, and the Compensatory Gestures of Children

I'm happy and I'm delighted and I fart and I laugh.

Aristophanes, *Peace*

For the Pennsylvanian—proud, earnest, at times almost inconceivably laconic—the family unit generally provides the informal boundary of "acceptable" flatus, with social norms defined not by elders† but by older adults of breeding age. These adults determine the degree to which flatulence is allowed within the family unit (audible public deflation is rarely intentional but,

---

* Some back-of-the-envelope calculations here for the Academy's astrophysicists. Assuming the flatus to be well outside of Earth's gravitational pull, while ignoring the motion of the planets, the minute but cumulative propulsive effects of solar radiation, the influence of microgravity, Lagrange points, etc., a single burst of human flatulence would, in about twenty thousand years, arrive on Mars.

† The role of elders should not be neglected in this respect, both in terms of the blasé pragmatism they embody and the embarrassment they occasionally engender in their descendants.[1] They are at once a model, to quote one subject, of "how not to give a fuck" and an object lesson in the dangers of subverting societal expectations. I fear it is quite possible that the ongoing shift toward nursing homes and late-in-life care facilities, and the ensuing removal of the elderly from the homestead, is drastically affecting the perception and role of flatulence in *all* age groups of Pennsylvanian culture.

[1] One female subject, for instance, a great aunt, aged eighty-five, Catholic, no longer bothered to depart her nightgown, instead daily positioning herself at a table in the kitchen and engaging with various family members in cutthroat bouts of dominoes, punctuating her genial table banter with the occasional exclamation, eyes raised to the heavens, "There goes an airplane."*

* The general incontinence and indiscretion of elders with dementia was a recurring theme of concern among the adolescents.

depending on the age and social position of the deflator, may be severely punished). These adults also determine the locales of acceptable family deflation—home, hotel room, movie theater, hike in the woods, family vehicle—and the procedures relating to it—opening a window, begging pardon, etc. There is great variation in how these parameters are defined.*

Born into this milieu, the Pennsylvanian Male is hardly different from any other newborn animal. Weak, defenseless, floppy, his activities are initially limited to crying, feeding, and evacuation. On occasion, he registers a feeling of distension around his middle, one that burgeons uncomfortably. And he learns that by drawing his elbows to his knees, the discomfort is relieved, accompanied by a light trumpeting at his far end. But he hardly registers these emanations as flatulence. Even well into his first or second year, his gaseous productions arouse little special attention from him. Lacking (1) the sphinctrous dexterity necessary to control his flatus, (2) the intellectual development required to initiate or sustain subterfuge, and (3) the historical framework necessary to place his actions in the context of larger society, he farts freely, frequently, and without remorse. He farts in pediatricians' offices and in mosques, at weddings and at funerals, in bathtubs, at restaurants, on conference calls. He may begin to find humor and entertainment

---

* In one crowded ranch home containing a family with three Adolescent Males, flatulence was so commonplace, infiltrating all aspects of daily life, that its appearance was hardly acknowledged, while in a dilapidated Victorian containing only a mother and her young son, a prodigy on the piano, the escape of a flatus mid-Liszt was treated as a sort of barbarian incursion.

in the *physical* sensation,* just as he finds humor in clapping or in falling down, but he has not yet understood flatulence as the markedly *social* endeavor it really is.

The first understanding of flatulence *as a signifier* comes usually between the ages of one and a half and four as he undergoes the process of potty training. Whereas previously his flatulence has elicited scattered adult commentary, now every flatus draws the concerted attention of his elders. During this phase, a fart is often greeted with the phrase "Do you need to poop?" or "Did you just poop?" And he is then carried quickly to the nearest toilet. Once atop the toilet (nearly always white porcelain, though sometimes a miniature plastic simulacrum) he is encouraged to empty the contents of his bowels. Any on-potty flatus earns him praise: "Good work!" and "Keep going!" He begins to notice now too that in certain times/places/company, his gaseous productions elicit different reactions from his parents, and so, even before the cultural status of flatus has fully gelled in his developing mind, the boy of Pennsylvania is already beginning to map his world into flatulent and nonflatulent zones.

A period of experimentation ensues. He passes flatulence in movie theaters, at parades, on swings, under tables, all to learn about the extent of his bodily domain. (Of equally great significance, the differentiation of scents by positive and nega-

---

* One three-year-old registered ecstatic disbelief at the bubbles surfacing in a swimming pool.

tive connotation occurs during this period. He begins to learn that certain things smell "good" while others smell "bad.") Though little concerted experimentation has been done upon this subject, it may be too that even in this early developmental stage, flatulence is already being in some sense "gendered" in Pennsylvania,* with male flatus receiving greater leeway or tolerance, thus laying the groundwork for the relatively distinct reactions to flatulence in males and females during preadolescence and beyond.†

A neural revolution has been building quietly during this time, and around the age of four, the boy makes a developmental leap: he begins to realize the world is full of discrete consciousnesses. With this new idea of the world, he develops the ability to actively modulate his own behavior to suit his environment. He becomes capable of presenting an exterior self that does not reflect his interior state; he is able, in other words, to lie and be polite.

Along with the ability to modulate his behavior comes a resentment in having to do so. And while the possible expressions of this resentment are limitless—he may scream at Arby's, run nude through a summer camp, throw snowballs at passing cars, draw penises upon public property—flatulence occupies a place

---

* The father, for instance, who, when his young son farted while climbing on a piece of playground equipment, said with a studied lack of affectation, "Gesundheit."
† Though it is also possible that a disparity in mature attitudes might result in part from other factors, the increased flatal volumes of males, for instance, or the higher degree of olfactory sensitivity among females. As with so much of our relations to one another as social beings, one yearns for a scientific control.

above all the rest. It is a subtle act of defiance, deriving its talismanic majesty from its ostensibly inadvertent nature and from the taboo against discussing it. Quite early then, well before the onset of puberty, the fart becomes a compensatory behavior, a way of claiming agency in situations where he might otherwise have none.[*] If he chafes at the strictures of public school, he might imitate the squelching sound of a fart when his teacher's back is turned. Or if he finds himself excluded by a would-be playmate, a fart sound may be deployed as a mode of ridicule when the boy squats down. At home, he may attempt to fart on parents or older siblings or other symbols of authority—a parent's phone, the television remote control. To him, nothing is funnier than a fart.[†]

And yet the more delight he takes in subverting parental authority, the more his own farts mortify him. While he may release flatulence at home, when he is in the society of other boys and girls, he is far less likely to employ actual farts to humorous effect. Instead, he is only a master of the pseudofart, always ready with one hand in his armpit[‡] or with his mouth nuzzled into the crook of his elbow. To be caught farting by his

---

[*] During a moment of social challenge, one male, aged four, called his mother "poo-poo," his father "diaper," and attempted to make flatulent sounds on the bodies of his peers.
[†] See *Ren and Stimpy*, *Dumb and Dumber*, *Lion King*, *Family Guy*, *Shrek*, Bart Simpson, *South Park*, *Step Brothers*, *Nutty Professor*, *Swiss Army Man*, *Caddy Shack*, *Rain Man*, *Austin Powers*, *Naked Gun*, *Last Action Hero*, *Ace Ventura*, *Blazing Saddles*. . . .
[‡] What choir of eight-year-olds, having suffered the direction of an undercompensated choirmaster for several months, has not had the quiet following a rendition of "Silent Night" broken by the fart-like squelch of a boy's hand squeezed in his own armpit?

schoolmates, or even falsely fingered as a farter, is possibly the most embarrassing thing he can imagine, and he and his male peers sometimes bicker away an entire afternoon assigning and evading blame for one fart or another. In fact, so socially disastrous is it to be caught farting that even to admit that one smells a real and actual fart in Pennsylvania is considered dangerous.[*] And so, through elementary school cafeterias, down mall escalators, and across soccer fields, the farts of little boys drift lonely and unclaimed.

3. Doorknobs and the Preadolescent Male

it was a bold tricke
to fart in the nose of the bodie pollitique

John Hoskyns,
"Censure of the Parliament Fart"

At the end of this period, sometime between the ages of eight and ten, the game of doorknob usually makes its appearance in the life of the Pennsylvanian Preadolescent Male. The rules of the game are as follows:

- In the company of his male peers, a boy may fart at any time.
- If he farts, he must immediately say "safety."

---

[*] A sort of schoolyard omertá is enforced via the maxim "He who smelt it, dealt it."

- If he fails to say "safety," any other boy may say "doorknob."
- Once "doorknob" has been called, all boys present are given license to punch the farter until he touches a doorknob.
- Once begun, the game never ends.

None of this is clear to the Pennsylvanian boy, of course. No one sits him down to explain the game, and there is no handbook or manual by which he might familiarize himself with its ins and outs. Instead, one hot summer day, five boys, ranging in age from eight to twelve, are standing around a basement deep freeze, divvying up ice-pops. Then someone, perhaps he himself, farts. Usually, such an action would elicit ridicule, but in this instance, one of the older boys yells, "Doorknob!" Before he knows what is happening, the farter is being pummeled from all sides. "You've got to touch a doorknob," one of the more considerate boys tells him, while punching him on the arm. He flees, still being walloped on his back, glancing this way and that in the basement gloom for the nearest knob. The moment his fingers graze the brass, the blows evaporate. Thus he learns, in a rough way, how to play the game that will dominate the next few years of his life.[*]

---

[*] As Bourke reports, it has been played in Pennsylvania, under various guises, since at least the nineteenth century, meaning it has outlasted two world wars, two pandemics, countless economic catastrophes, and the invention of the automobile, the atomic bomb, and the Nintendo.

One is struck initially, as is the Pennsylvanian himself, by the violence of this game. By eight, the Pennsylvanian boy can whistle, sing, run, jump, climb a tree, shimmy between the gates of a chain-link fence, pat his head while rubbing his stomach, expel gas while walking without a hitch in his step. But for all his bodily ease, he receives little physical contact from others. Without yet having attained anything like a romantic life, he has left behind the caresses of his parents. Into this vacuum of physical touch comes the game of doorknob. And though the game may seem cruel, there is little he treasures more than the violent physical attentions of his fellow boys. In fact, when a doorknob is not too distant, he will refrain from saying "safety" for no other reason than to experience the thrill, camaraderie, and pain of the doorknob chase.*

For a time, the Pennsylvanian boy is understandably consumed by this game. Wherever he goes, he takes note of the available doors, studding the internal map of his environs with doorknobs.† The very feeling of laying his hand on a cool knob fills him with reassurance.‡ He overlaps doorknob with other

---

* As the Harrisburg folklorist Trevor J. Blank reported: "One boy proudly stated, 'the game is best when you're out camping or somewhere where it's hard to find one to touch!'"
† One man, aged thirty-six, recalled that for a handful of years during his teens, he never entered a space without acquainting himself with all available doorknobs, a form of conditioning so strong that even today, when he feels the urge to fart, whether in company or alone, he sometimes unconsciously scouts for nearby doors.
‡ Importantly, he is utterly incognizant of the sexual undertones of such extensive knob-rubbing.

games—staring contests, laughing contests, Marco Polo—to see which rules take priority. In the company of his peers, he farts wherever and whenever he feasibly can, experiencing the double release, of gastric pressure on the one hand and of social pressure on the other. And he and his comrades evaluate one another's flatulence with theatrical gagging and nose pinching, with certain farts and farters attaining a legendary status amongst them.* † As his credo, he takes a line from Athenaeus's *The Learned Banqueters*: λαβὼν ἕκαστον αὐτῶν κατὰ μέρος προσπαρδέτ.‡ If he should pass gas while playing video games alone, he may murmur a soft, subconscious "safety."

My more astute colleagues will recognize immediately the revolution in flatulent values this game represents. Not only is the boy now permitted to deflate outside his family unit; he is encouraged to do so. He throws off the taboo of earlier years and its incumbent embarrassments, for the first time since infancy acknowledging his flatulence and claiming its smell as

---

* See the cutting-edge research being done at ICEF in Montreal.
† See the likely medieval tale, related by Bourke, of a knight in Cornwall who "having gathered together in one open market-place a great assemblie of knights, squires, gentlemen, and yeomen, and whilest they stood expecting to heare some discourse or speach to proceed from him, he, in a foolish manner (not without laughter), began to use a thousand jestures, turning his eyes this way and then that way, seeming alwayes as though presently he would have begun to speake, and at last, fetching a deepe sigh, with a grunt like a hogge, he let a beastly loud fart, and tould them that the occasion of this calling them together was to no other end but that so noble a fart might be honoured with so noble a company as there was."
‡ "Take these things and fart upon each of them."

his own.* In so doing, he and his compatriots claim some small cultural territory for themselves and their own fledgling male society.† It can be no surprise that flatulence, which he formerly employed as a sort of asymmetrical warfare against the cultural authority structure, should be reborn as the bedrock of this new social group.

Sadly, as the boy progresses into and through adolescence, the game of doorknob fades away even as the flatulence remains. While initially a means of negotiating with the power structures of the larger culture, deflation quickly becomes part of the power structure of his own Adolescent Male society. Flatulence might be employed to end a discussion, thereby quashing dissent. Or it might be used to exclude participants from certain places and activities, as in "Horatio and Devon were farting up the sound booth, so we left." The brazenness

---

* Consider, for instance, Mike Myers, glossing Marston's "Every man's turd smells well in's own nose," altering Florio's translation of Montaigne's "*Stercus cuique suum bene olet*," itself a gloss of Erasmus's Adagia, "*Suus cuique crepitus bene olet*," which is itself a translation of Apostolius' *Paroemiae* Έχαστος αὐτοῦ τὸ βδέμα μήλου γλύκιον ἡγεῖται, "everyone thinks his own fart smells sweet," which is mentioned, as well, by Burton, in a footnote to the Wazir's accusation, "every time thou fizzlest, thou smellest and sniffest at thy fizzlings," in the tale of the "Four Hundred and Eighth Night."

† It should be noted, however, that this custom is largely limited to peers of the same gender, and more infrequently to peer groups of differing genders. Less commonly does he fart audibly amongst elders outside of his nuclear, or in some cases extended, family unit. And very rarely does the adolescent male willfully fart audibly in unfamiliar social contexts (an example of this anxiety: one comedian, performing in Pennsylvania, advocated the practice of releasing a "test balloon," a silent fragment, to ascertain precisely how malodorous, and thus offensive, the complete fart might be). Considering the limited context for the public fart, it must be acknowledged that this gastrointestinal show of force, for all intents and purposes, is exactly that: a show, a performance for one's peers, signifying the more audacious and emancipated social role to which one now aspires.

of a boy's flatulence (i.e., his willingness to deflate in unfamiliar company or locales) may even provide a barometer for his current position in the adolescent social hierarchy, and thus a measure of his suitability as a potential mate. One Pennsylvanian recalled an occasion when a number of Adolescent Males were competing for the attentions of a single female: "There were four of us. We were all sitting around in her bedroom. My cousin, Joey, was the oldest. He was sitting on the bed with her, and he had his arm around her. We were talking, then Joey cut everybody off. 'Hold up, Ima bust ass.' He leaned into the girl and let a fart come out the other side. She seemed angry that he farted on her bed, but that night they hooked up."

It must be noted, however, that just as the game of doorknob has no official beginning, neither does it ever officially end. One man in his thirties recalled that on a recent backpacking trip someone had called doorknob on him, "days from the next encounter with a doorknob." No matter how far he travels, in space or in time, he never fully escapes that period of his life when every act of flatulence carries with it an intimation of fellowship.

Intimacy

A bedouin in Yemen goes to live in town and becomes a wealthy merchant. When his first wife dies, he wants to marry again. At a solemn moment during the wedding, however, he cannot help letting go a loud fart. Humiliated, he runs away from the wedding, leaves his hometown, and goes to live in India. Ten years later he returns to his native town, convinced that everyone must have forgotten about him and his abominable fart. Sneaking around, he overhears a girl ask her mother when she was born, and the mother answers: "You were born on the day that Abû Hasan farted." Shocked by the experience of seeing the date fixed according to his mishap, Abû Hasan returns to India and dies in exile.[*]

Sir Richard Francis Burton,
"How Abû Hasan Brake Wind"

On the topic of intimacy, one brief final comment is necessary on the flatulent rituals of Pennsylvanian Males. As the male progresses falteringly out of adolescence and into the dyadic stage of sexual pair-bonding, the customs of homosocial flatulence gradually[†] erode. The concerns of youth recede, along with the company in which those concerns found expression. In their place the peculiar habits of adulthood take root. The thirteen-year-old who scrabbled for the nearest doorknob becomes the thirty-year-old with a life partner, a hybrid sedan, and a taste for natural-process pour-over coffee. In part, this alteration is purely circumstantial in nature, relating to the process of "settling down" with another human being and the ensuing shift in living arrangements and daily schedules. But

---

[*] A flatulent abhorrence similar to that reported by Edward de Vere: "This Earle of Oxford, making of his low obeisance to Queen Elizabeth, happened to let a Fart, at which he was so abashed and ashamed that he went to Travell, seven yeares. On his returne the Queen welcomed him home, and sayd, My Lord, I had forgott the Fart."

[†] The process is markedly protracted in some.

we would be remiss to think of this shift as an end to the Pennsylvanian Male's flatulent customs—flatulence, after all, does not go away simply because one ceases to be a teenager in Pennsylvania. Rather this shift represents an *evolution* of those customs.

It can scarcely be credited to coincidence, after all, that the period of male social flatulence should occur in concert with the maturation of an immature child into a sexually functional adult. Whatever its other roles—whether as a buttressing of hegemonic masculinity, a psychoprotective cocoon[*] for the metamorphosing male self, an unsubtle form of sexual competition—in its most basic function, the highly ritualized flatulence of childhood prepares the young Pennsylvanian Male for a different form of flatus in adulthood, the flatus that, almost by necessity, accompanies any intimate sexual relationship.

Indeed, one hardly needs to look far into the historical record to see the psychological pressures associated with intimate flatulence. In the thirty-fourth tale of the *Uji shūi monogata*, when a woman deflates in the middle of a moonlit tryst, her partner flees, determined to become a monk. In Chaucer, the cuckold receives not a kiss on his brow but a fart, and Abû Hasan elects exile above the comforts of a home where he is known to deflate.

---

[*] Not entirely dissimilar from the way certain academic treatises cocoon themselves in an off-putting miasma of footnotes.

At some moment in his adulthood the Pennsylvanian Male will find himself enthralled by the company of another human being. Seated on a futon, one hour of conversation will turn to two, two to four, until it's the next morning and they both lie spooled in each other's arms beneath a too-thin comforter from Target. All the while, he will have restrained himself from passing gas, unconsciously at first, then with great effort, until the sweat beads on his forehead and the air can be heard grumbling in his midsection. For weeks, possibly even months, their budding relationship will follow this same rubric: a series of sexual and intellectual encounters during which no one ever really farts. Then one morning, waking to the sound of the covers rustling, he will roll over in bed and find his partner cautiously venting the duvet. What can it be that enters the room then but love?

Love, that plaything of poets; love, whose fundamental essence is forgiveness; love, that "country whose violence remains alien and overwhelming." It is to this country that, like a passport, one's pungent bodily sulfides permit entrance. For the Pennsylvanian, to share a bed is, eventually, to submit oneself to the flatulence of another and, more importantly, to submit one's flatulence to the judgment of another. And it is not the exchange of vows or the designation of beneficiaries or even the gasping elisions of sexual congress, but flatulence that is often the original, and sometimes the ultimate, token of Pennsylvanian intimacy. It is flatulence, frequent and unremarkable flatulence, that separates the flitting sparrow-like romance of tweens from the wizened love of old age.* In this respect, it is difficult to see the crude knock-knocks of the Pennsylvanian Male, the whoopee cushions, the gleeful doorknob sprints, as anything but a preparation for that moment when, entwined in the limbs of another human being, his bowels begin to rumble.

---

* Primo Levi recalled his aged Aunt Regina and Uncle David sitting together at the Café Florio on Via Po, the former scolding the latter, *"Davidin, bat la cana, c'as sento nen le rukhod!"* In *rukhod*, of course, my colleagues will have already recognized the plural of *rúakh* (a transliteration of the Hebrew חוּר), translated variously as "breath" or "spirit," but meaning in its most direct sense "wind." "David, thump your cane, so they don't hear your winds!" Though there is a sense here of exasperation, of the frustrated cul-de-sacs of marital life, I am more struck, as one who will always now in some sense be Pennsylvanian, by the care and familiarity of the supplication, by the lifetime of shared flatulence upon which such moments are built. Isn't this intimacy, this willingness of one human to acknowledge *the whole* of another and the consent to be so acknowledged in turn, isn't this more than a little miraculous? The conjugation of humans is so common, we forget what enduring bravery such a covenant requires.

There is yet another meaning to Aunt Regina's plea, as well, as Levi notes, *rúakh* being better known for its beguiling use in the Masoretic text and thus its place in the second verse of Genesis:

ה יָנְפּ לַע תֶפֶחְרַמ םיהֹלֱאֶ חוּרְו

"And a rukhod of Elohim blew across the face of Tehom," or, as the King James prefers it, "And the spirit of God hovered over the face of the deep." To call the flatulence of a lover *rukhod*, then, is to offer a far less spiritual gloss of our beginnings, a sly sidelong Midrash, a narrative where the face of one god is brushed by the *rúakh* of another, and only then is creation made possible.

VIII

Breath

Our bodies are not up to us.

Epictetus

Terms

n. exhalation, expiration, wheeze, pant, *nephesh*

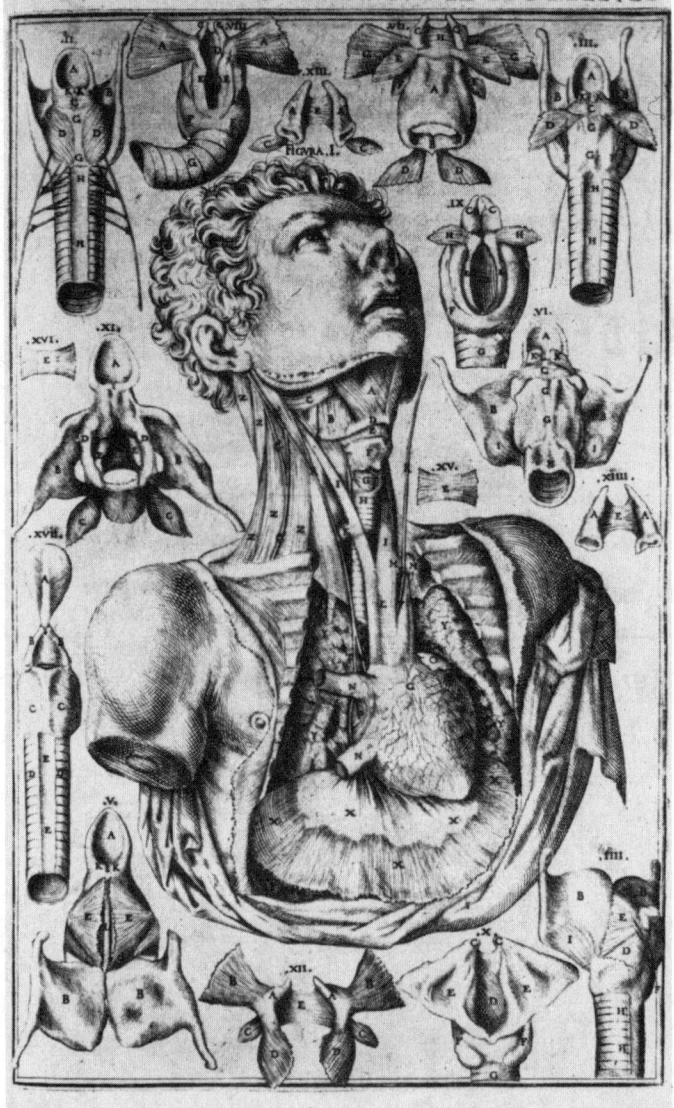

Anatomical view of the heart, lungs, throat muscles, and various related internal body parts, *De vocis auditusque organis historia anatomica singulari fide methodo, tabula XIII,* Giulio Casserio, 1601

Biological Prologue: The Act of Breathing

The act of breathing will be more or less familiar to most readers. The eyes flutter closed, the nostrils flare imperceptibly, and from the pre-Bötzinger complex in the brainstem, an electrical impulse skitters down ion channels in the phrenic nerve to the C4 junction. The diaphragm contracts, flattening downward against the stomach and intestines, while the intercostals tighten between each rib, drawing the rib cage up and out. The abdomen and chest expand like a bellows, increasing the volume within the thoracic cavity and creating a pressure differential between the inside and the outside of the body. The atmosphere, which has been bearing down all along at about fourteen pounds per square inch (sea level), rushes in, seeking equilibrium. If the mouth is closed, there is only one means of ingress.

With respect to breathing, one is tempted to call the nose central, perhaps because of the role it plays as a sort of grooming station for the raw incoming air, perhaps because of its position on the face. The air from the surrounding atmosphere is drawn in here through the nostrils, usually two in number. It crosses the nasal vestibule and is funneled on each side between three conch-shaped shelves of flesh and bone into four horizontal passages. The curved shapes of these shelves swirl the air as it passes through—cleaning, warming, and humidifying it in the process. Thus primped and preened, the air is drawn upward into the olfactory cleft and offered to the receptor cells cloistered there, where, through the binding of patterns of molecules, the brain identifies scents—coffee, wool, flatulence, Pantene. At this point, nearly ready to become breath, the air rushes down toward the throat, at the very last moment receiving from the paranasal sinuses a spritz of the vasodilator nitric oxide.

Through the ringed cartilaginous trachea, through the branching of the bronchial tubes, the air goes down, down finally into a pair of lobed, asymmetrical lumps of sponge and sinew.

In humans, the lung is an organ for the manipulation of surface area. Its job is to put as much air as possible in contact with as much blood as possible, and its solution is the alveolus. The architecture of these alveoli presents some descriptive difficulties in a basic primer on human breath, in part because they're so small (about two hundred micrometers across) and in part because the human brain, despite its capacity for reflection, argument, memory, and deception, does not quite possess the three-dimensional processing power necessary to render up a mental image of uncounted millions of interconnected elastic polyhedrons grouped in sacs and suspended in a mess of sinew, mucus, and threadlike capillaries. But despite their minuteness and peculiar structure, the alveoli can hardly be overlooked if one wants to understand breaths, since it is here that breathing happens.

Though the alveolar sac is often depicted as a grapelike cluster, with each grape being a single alveolus, the alveoli are not spherical like grapes. They are geometric in nature, with soft-edged planar surfaces, more akin to the faceted arils of the pomegranate. Their walls are very thin, covered on the inside with a surfactant (to help the structure withstand the pressures of breathing) and netted on their outside in a skein of capillaries. As we mature, minuscule passages—the pores of Kohn, the channels of Lambert—form between neighboring alveoli and between alveoli and nearby bronchioles, meaning they are hardly even like the discrete arils of a pomegranate and more like "pronounced surface irregularities with openings."

The alveoli in human lungs are so tiny, so delicate, and so numerous that they've never actually been counted. The only way to arrive at an idea of the number of alveoli is to take a small piece of lung, osmicate it, embed it in a fixative like glycol methacrylate, cut it into thin slices, and try to extrapolate some notion of the three-dimensional whole from these two-dimensional sections using the Euler number. But even this relatively rudimentary approach is challenging. The structures of the lungs aren't homogenous, meaning that two given samples of the same lung might be wholly different, or the same sample might produce different results if sliced along a different axis, and any projections based upon these slight differences will then produce wildly divergent estimates. It is a little like looking at several oak leaves cut into strips and being asked to estimate how many leaves the tree contains, if the tree is eight hundred feet tall and made of gelatin. Despite these challenges, stereologists have been able to come to some idea of the number of alveoli. They estimate there are around half a billion alveoli in the average human. With every breath, these pronounced surface irregularities stretch ever so slightly, in the process spreading a little more than a soda can of air across a pool of blood the size of a tennis court.

Oxygen, of course, is the reason the whole apparatus exists. In the form of two atoms bound together as a molecule of O_2, it enters the alveolus. There it is drawn through the alveolar wall into the bloodstream and saddled atop red blood cells on proteins of hemoglobin. Carried in the pulsing current, it rides these cells into the left side of the heart and out into the ever-diminishing corridors of the vascular system, to the brain, the liver, the tip of the big toe, where it is used by cells to break down packets of chemical energy. The result of this process—carbon dioxide—is then brought back by the blood to the lungs along with other wastes. With the relaxation of the diaphragm and intercostals, elastic fibers in the lungs then return the thoracic cavity to its resting form, and the air is donated once more to the atmosphere.

This, then, is a breath. Average duration in the mature human: five seconds.

By following the above procedure, a human being will produce roughly 20,000 of these on any given day, 600,000,000 in a lifetime. It is possible to identify a breath without advanced medical equipment or training: simply place a finger horizontally beneath the nasal orifices and wait. The last breath can be easily distinguished by the absence of further inhalations.

A Manual for Collection

Though some small technical challenges exist, relating mainly to the capture and storage of low-molar gases, those interested in breath collection will find it a low-cost and approachable undertaking. It requires neither extensive financial outlays nor esoteric knowledge. Moreover, most households have on hand the basic materials necessary to begin a collection: with little more than a jar, a lid, and some rubber bands, almost anyone can become a collector. It is only a matter of finding the appropriate breath.

The selection of a collectee will be the first consideration. While the pool of available candidates is limited only by considerations of geography and consent, the novice is well advised to make their maiden collection from a close personal friend or family member. The reasons for this are two. First, it offers the opportunity to hone one's craft in a familiar and controlled environment. Second, it reduces travel time.

Oxygen Nation

The early atmosphere was composed of simple hydrides such as water vapor, methane, and ammonia. Outgassing of the still forming planet added nitrogen and carbon dioxide to the mixture. Only two and a half billion years ago did a ray of light strike a blue-green bacterium, converting six molecules each of carbon dioxide and water into a molecule of sugar. The byproduct of this process was an element so biologically dangerous that the bacterium immediately evacuated it: oxygen.

A collectee having been selected, attention is turned to the matter of the collection apparatus, which is itself necessarily determined by the type of breath to be collected. Countless potential breaths present themselves continuously to the would-be collector, from hyperventilated cries to apneic snorts, each one a legitimate object for study and reflection, but this very abundance of breath requires a great deal of selectivity. To the serious collector, the object of primary interest will be that exhalation which, failing to be followed by further inhalations, reveals itself as the last.

Tip

When selecting a collectee, it is important, for practical reasons, to take into account the expected date of collection. Ideal candidates will be projected to produce a last breath within three to six months, but will still exhibit the compos mentis [soundness of mind] required to complete all documentation. Those diagnosed with late-stage nonpulmonary cancers are excellent candidates. A wallet-size list of recommended illnesses may be found at the back of this pamphlet, along with revocable and irrevocable contract templates.

In considering how best to capture and contain a last breath, students of the history of breath collection will necessarily be drawn to the methods by which Joseph Priestley collected the last breaths of mice—a bell jar set into mercury, housing a platform on which a still-breathing mouse was placed—by which means the presence of carbon dioxide was deduced. But while such an apparatus ensures uncontaminated specimens, these methods are impracticable for collecting from humans, owing to the effects of limiting the collectee's access to oxygen and the difficulty of obtaining suitably sized bell jars. Those seeking a "low impact" approach may wish to consider the method used to collect Thomas Edison's last breath—test tubes left open at the bedside and later capped with wax. However the steady mixing of gases in a room, combined with the rapid effusion of gas through wax, will yield a highly diluted final product. For those still new to the work of breath collection, it is often advisable to work with whatever materials are most familiar, focusing instead on honing the techniques of breath collection. A glass jar or bottle, handled well and capped quickly, will serve better than more complex and unfamiliar apparatuses.

The active period of collection commences when, the collection vessel having been procured and a memorandum of understanding having been signed by all parties and notarized in the presence of two witnesses, the collectee embarks, in any of the many possible manners, upon the process of dying. Though death has long been determined by cessation of breathing, there is some uncertainty here regarding when precisely dying begins. "On the Definition and Criterion of Death," from *The Annals of Internal Medicine,* understands death "not as a process, but as the event that separates the process of dying from the process of disintegration." It makes no distinction, however, as to the event that separates the process of living from the process of dying. Neither does it provide guidance as to the criteria upon which one might recognize such an event. Frequent sighs—deep arrhythmic exhalations—may suggest that the process of dying has begun, but these cannot be treated as reliable indicators. Given this lack of clear benchmarks and criteria, it is advisable to remain in the collectee's company continuously in the run-up to collection.

Deep Breaths

Though whales are able to collapse their lungs while diving, humans never fully exhale. The collective collapse of the alveoli in most mammals is a catastrophic physiological event. When individual alveoli collapse in the course of normal breathing, they must be reinflated. To do so, a breath is taken, and then another larger breath. Known as a sigh, these reinflations occur in humans roughly once every five minutes (more frequently in smaller mammals). While sighing as an emotional activity remains poorly understood, on a neurological level, there is little doubt that the sigh is essential to survival. Mice perish almost immediately when engineered without the ability to sigh.

Now is the time to become familiar with the collectee's patterns of breath. By discreetly synchronizing their own breaths to the inhalations and exhalations of the collectee, the collector will begin to develop an understanding of the pace, depth, and feeling of the collectee's respiration. If helpful, the collector may recall the training of the monks of the Heian and Kamakura periods who, believing that one's release from the cycle of rebirth hinged upon the arrangement of the soul at the exact moment of death, would mirror the breathing rhythm of a terminally ill peer, with every exhalation reciting a prayer to the Buddha in unison with the dying. Following the instructions set forth in *Ichigo taiyō himitsu shū*, the collector may choose to dwell within this rhythm, "one day, two days, seven days," or may even continue the practice, as the monks did, "until the breath ceases."

Though it is unclear whether the purity of the final exhalation can save one the insult of being born once more into this world of striving and suffering, as the monks believed, a familiarity with the collectee's respiratory rhythms may aid in collection. There is evidence that the electrical impulses governing breath impact the workings of the brain and nervous system as a whole. The regular jolt not only gives a rhythm to the pulsing neuronal activity in the limbic system but also strengthens signals being sent out to the body. By mirroring these respiratory rhythms, the collector will coach their own organism nearer to the mental and physical state of the collectee, tracing the arc of cerebral function loss that will soon leave the brain with only its most basic prerogative: to breathe.

OXYGEN NATION

THOUGH OXYGEN IS CENTRAL TO OUR EXISTENCE ON THIS PLANET, THE INTRODUCTION OF SUCH A HIGHLY reactive element to the atmosphere is believed to have eliminated nearly all life then present, causing Earth's first mass extinction. The advent of photosynthesis, for this reason, is sometimes known as the Oxygen Holocaust.

Moments of Inspiration

Feuerluft, or "fire-air," was the name given by pharmacist Carl Wilhelm Scheele in the early 1770s to a kind of air, derived by heating manganese oxide with sulfuric acid, that sustained the burning of a candle. A year later, in a series of experiments involving alternately placing mice and mint in "a quantity of air, made thoroughly noxious, by mice breathing and dying in it," Joseph Priestley solved what he believed to be one of the most pressing philosophical questions of his day—"The provision in nature for restoring air, which has been injured by the respiration in animals"—in the process grasping what he, in accordance with a misguided theory of combustion, would come to call dephlogisticated air. But it was Antoine Lavoisier, synthesizing and systematizing the work of Priestley and Scheele, who first realized that an irreducible substance existed in air, naming it, in accordance with a misguided theory of acids, by the Greek for "giver of acid": oxygen.

As the moment of collection approaches, its trajectory can be traced in a journal set aside for this purpose. Here the collector should make note of any changes in habit and appearance (sallow skin, a disinterest in lengthy novels). Little by little, much of the body's usual functionality will now begin to subside. Appetite and thirst will decrease. Periods of sleep will become longer, and mental lethargy and bodily restlessness will encroach upon those remaining episodes of wakefulness. The collectee will find it increasingly difficult to evacuate all the materials they once handled with ease. The movements of the bowel will slow. Urine, concentrated from dehydration, will appear resinous in color and smell. As saliva and mucus gather in the throat, the collectee will cough or gurgle. If it is difficult to treat these final emanations without repugnance, the collector should recall that Siddhartha Gautama was said to have washed the sheets of a monk dying of dysentery, believing that even the most earthly materials offered the opportunity for compassion.

To alleviate discomfort, cool damp cloths may be placed over the forehead and eyes. The feet can be elevated. If the ability to drink has been entirely lost, the mouth may be moistened with an ice cube wrapped in a thin cotton towel or with small spoonfuls of chipped ice. Likewise, a balm may be applied to the lips to prevent chapping. Pain may be treated with prompt subcutaneous injections of morphine. If the act of breathing is itself a seat of anxiety, benzodiazepines may help to ensure that the respiratory rhythm is not perturbed unnecessarily. A bedside table will serve well to hold medications, syringes, collection vessels, and tissues. It is recommended, at this point, that the collectee's hand be held firmly but gently. The gesture will be familiar from crossing the street.

Oxygen Nation

Due to the arrangement of its electrons, oxygen is a uniquely reactive element, ready to bond with a vast array of molecules. While this molecular promiscuity makes oxygen one of the body's most versatile and vital building blocks, it also means that, left unchecked, oxygen will quickly destroy an organism. Indeed, the body can only host the element by maintaining a suite of reactions that convert it into less reactive forms, systems that are quickly swamped by a surplus of oxygen. At higher-than-normal pressures, pure oxygen kills in minutes.

During periods of consciousness, the collectee may also experience mental anguish as the very preciousness of life begins to preclude its enjoyment. The respiratory rhythm may be interrupted by weeping. In these circumstances, the collector should recount humorous anecdotes, the time they were trapped together in a canoe during a thunderstorm. Special care should be taken to avoid stories that mention estranged relatives, ripe fruits, cool breezes, or the stifled giggles of children hiding beneath blankets. Reading aloud is recommended, though never an overly dense or serious work. *Charlotte's Web* works well, but one must be careful not to make it to the end of the fair.

Depending upon the nature of the mortal event, the sequence immediately preceding collection may last as little as a few minutes or as long as a few days. Some collectees may cling to life as though to a root on a washed-out riverbank, while others will depart speedily. (The forward-thinking collector will be sure to have packed snacks.) During this period, it is important that the hand of the collectee be held at all times, firmly but gently, as Charles A. Leale is said to have done for Lincoln at Ford's Theater, after performing mouth-to-mouth resuscitation.

As the brain retrenches, the autonomic nervous system will come to the fore, and the body will be animated by the automatic and involuntary rhythms of the smooth muscles. These processes, in turn, will not fail all at once but decay in piecemeal fashion, as an orchestra without a conductor falls unsteadily out of tune and time, the body resolving itself into an increasingly uncoordinated assemblage of faltering biological processes, until at last it can sustain only that function most fundamental to the organism's survival: respiration. Now, like a deranged tympanist, the brainstem obstinately strikes the diaphragm, ventilating the lungs with a ragged breath that serves less to further life than to bring up evidence of its progressive destruction.

Perhaps no more harrowing a moment attends the collection of a breath. At some moment, the collector will have reflected upon the belief that the soul flees the body with the final breath. James George Frazer's *The Golden Bough* may come to mind, for instance, with its descriptions of the Nias, binding the mouths of the recently deceased and bunging up the nostrils, to prevent the soul from escaping and taking up residence in those nearby, or of the Nisga'a physician who, accidentally inhaling a patient's soul as he leans over the sickbed, must be thumped on the back by his colleagues until he coughs it up. The collector may recall the hollow bones Frazer claimed Haida medicine men always carried on their person, for capturing departing souls and restoring them to their owners, or the snares set by the sorcerers of Danger Island, larger for fat souls, smaller for thin. But now, the collector seems to see in the body's insistence upon breathing not the fluent melancholy of a soul but a rat, desperately treading water. It is easy, at this juncture, to question the very enterprise of breath collection. To what end do we spend such long hours at a bedside? Why do we hold this hand? Often in the evenings, before consciousness and unconsciousness collapsed into a single state, the collector will have read aloud, going on reading even after the collectee fell asleep. Why? These and a thousand other questions will cross the mind, interspersed with memories of that period before the thought of collection had formed, when it was possible to sit together in the garden, talking about nothing in particular, without paying attention to each passing breath.

Atmos-Facts

GIVEN THE TOTAL NUMBER OF MOLECULES IN THE ATMOSPHERE ($\sim 10^{44}$), THE ESTIMATED NUMBER OF HUMANS TO HAVE EVER DIED ($\sim 10^{11}$), AND THE NUMBER OF MOLECULES IN EACH BREATH ($\sim 10^{22}$), ALONG WITH SOME FURTHER ASSUMPTIONS ABOUT ATMOSPHERIC

CIRCULATION, EVERY BREATH WE TAKE CONTAINS 100
BILLION MOLECULES OF LAST BREATH.

Slowly the breath will grow fainter. The eye having been trained by habit on the rise and fall of the chest, the ordinary second-by-second passing of time will most likely be usurped by the strange and unlikely rhythms of terminal respiration. Now the collectee pants, now great swathes of time pass without sign of life. Hours may pass in this state of suspension, or perhaps only minutes, it is often impossible to say.

By now, the collector will understand that the collection of a last breath is, more than anything, an exercise in ever subtler degrees of uncertainty, the principal source of this uncertainty being the determination of time of death. In fact, the final period of respiration is often so faint that it was not uncommon in the past, out of fear of error, to delay burial until evidence of putrefaction was observed. In the Leichenhausen of Munich, for instance, those deemed dead were arranged in ranks on glass biers, a bell connected by wires to a brass ring around their fingers, lest a reawakening go unnoticed.

The hands will now be slick with sweat. The back may ache. The eyes, when closed, will often reveal the drawn yet familiar face of the collectee, engraved on the retinas in wavering blue. As the breaths become few and fainter, the collector regularly uncaps the vessel and captures a breath, feeling, as the intervals between breaths widen, that each must certainly be the last, only to detect, in the slightest rustling of the bedspread, evidence of another contraction of the diaphragm. At last, an exhalation of indescribable faintness arrives. The collector reaches out, captures it, then watches the collectee closely, almost suspiciously.

Deep Breaths

EACH DEATH HAS ITS OWN RESPIRATORY SIGNATURE. THERE ARE THE DEEP KUSSMAULIAN GASPS OF THE DIABETIC UNDERGOING KETOACIDOSIS AND THE CHEYNE-STOKES CRESCENDO-DIMINUENDO CYCLING OF THOSE DYING OF CARBON MONOXIDE POISONING, AS WELL AS AGONAL BREATHING, ATAXIC BREATHING, AND BIOT RESPIRATION. A SENSE OF THESE TERMINAL RHYTHMS MAY OFFER CLUES AS TO THE IMMINENCE OF THE FINAL BREATH.

Finally, several minutes having passed, the collector may conclude the final breath has been captured. Only now does the body arch up from the bed, nostrils dilated, lips quivering, to gorge itself on one final gulp of air. Everything hinges upon what happens next. Firmly but gently, one hand squeezes the hand of the collectee while the thumb of the other quickly uncaps the vessel and positions it anterior to the lips. (Some manual dexterity is required for this maneuver, as the need to hold the hand complicates the uncapping; the collector will want to have spent no fewer than ten years in piano lessons.) Now, as the elastic action of the lungs tightens the chest, the final breath rushes out into the waiting bottle. It is only a matter of recapping the vessel to prevent its escape.

Respiration Station

During voluntary exhalation, activation of the primary motor cortex, the premotor cortex, and the supplementary motor areas enhance the activity of nonrespiratory muscles, causing the fingers to operate with greater peak force. Thus, matching one's exhalations to those of the collectee should ensure the vessel is re-capped with little difficulty. Those mirroring the collectee's respiratory rhythms, whether consciously or otherwise, will want to be sure at this juncture to inhale.

As with any work into which we throw ourselves fully, a certain melancholy attends the moment of collection. The vessel is so small, and it takes so much to fill. The collector will be forgiven if their thoughts, fleeing the present, turn to those who understood so little of what they were seeking but sought it anyway, to Scheele, dead at forty-three, poisoned by his own experiments, or to Priestley, fleeing England as a mob burns his library, or to the splendid head of Lavoisier, blinking up at Lagrange, so the story goes, from the basket of the guillotine.

Deep Breaths

While we smell during inhalation, it is our exhalations that circulate scents from the mouth up into the nasal passages to the

EARTHLY MATERIALS

OLFACTORY SENSORY NEURONS RESPONSIBLE FOR CREATING THE SENSATION OF FLAVOR: THE LAST BREATH IS THE LAST TASTE OF LIFE.

With the capping of the bottle, the main thrust of the work has been accomplished. Labeled with the name of the collectee and the time of death, the breath is now ready to be entered into a collection. At this stage, the face of the collectee, traditionally, is searched. The forehead may be kissed. The hand should be held until the undertaker's arrival.

IX

Feces

"Where do they go to the bathroom?"

Ramona Quimby, referencing the
characters in books

Terms

n. shit, scheisse, dookie, crap, gruntie, grumpie, caca, the runs, the Hershey squirts, dirty squirties, number two, the big number two, a deuce, diarrhea, turds, a dump, a smash, a floater (lighter-than-water feces), a sinker (denser-than-water feces), a deposit, a blowout (feces that escapes containment by a diaper and/or pants), a skid mark (feces on underwear or the indelible mark where such feces once was), turdlet, turtlehead (grogan in Australia; a lozenge of feces protruding unbroken from the anus), manure, a log, soup, rice water, poop, poo, poopy, poo-poo, a brick, a loaf, a load, pay dirt, dung, doo, doo-doo, doody, BM, bomb, chocolate, Montezuma's revenge, pebbles, coprolite (fossilized feces), ambergris (pearlized sperm whale fecal ejecta), scat, shart (a portmanteau of *shit* and *fart*), a healthy, a movement, stool, excrement, the paperwork

It's important to know what healthy poo looks like.

NHS

Share this chart with the people you care for to help them identify whether they may be experiencing constipation.

Type 1

Separate hard lumps, like nuts (hard to pass)

Type 2

Sausage-shaped but lumpy

Type 3

Like a sausage but with cracks on the surface

Type 4

Like a sausage or snake, smooth and soft

Type 5

Soft blobs with clear-cut edges

Type 6

Fluffy pieces with ragged edges, a mushy poo

Type 7

Watery, no solid pieces. Entirely liquid

If a poo does not look like type 3 or type 4 it could be constipation. Contact the GP surgery of the person you are caring for.

Diagnostic stool scale, Bristol Stool Chart, NHS, 2023

Biological Prologue: The Tube

In very general topological terms, the human body is a tube. At one end of the tube is a mouth, at the other an anus, and everything else—the hands, the liver, the brain—is just window dressing. We are all variations, by this thinking, on the theme of earthworm. It's assumed the tube evolved from the simpler "blind gut" digestive sacs of animals like the sponge, but whether the original digestive orifice was a mouth, an anus, or both remains a matter of heavy debate. While we think our insides are part of our organism and thus under our control, in reality, the outside world pierces us as simply as a spear. This is why the cells of our gut share an evolutionary lineage with those of our skin; they're really just another way of interfacing with that outside world. The tube's great innovation is that it gives us a surface where those interactions can occur in a selective and controlled manner. Unlike the oak tree, absorbing whatever its roots touch, we get to choose what comes into contact with this surface and when. The tube itself does everything else, fostering an organized series of interactions that provide energy to our organism. Which is to say that when you get right down to it, we really have just one job on this planet: get stuff in one end of the tube and let it out the other.

The process is relatively straightforward. Chewed food travels down the esophagus to the stomach. After being churned there into a pulpy acidic fluid known as chyme, it passes through the pyloric sphincter into the twists and turns of the small intestine. Here the matter is slowly mixed and carried forward by rhythmic phasic contractions (RPCs), involuntary and insensible squeezes of sequential rings of muscle similar to the peristalsis by which the earthworm moves. As it passes into the large intestine, or colon, it moves first up the right side of the abdomen, then laterally across the body just below the ribs, before descending down the left side to the rectum. (A doctor-friend palpates clockwise along this route to motivate the bowels of her constipated patients; she also uses one of those purpose-built toilet step stools in her home to raise her knees to what she believes is the ideal height for maximum ease of evacuation.) During the colonic portion of the passage, water is drawn from the mixture, and bacterial fermentation allows for the digestion of some dietary fibers. At this point, the matter is prepared to pass into the rectum. Held back mainly by two complementary sets of sphincters, an internal involuntary sphincter and an external voluntary one, it sits here and waits.

When another meal is eaten, the sensation of fullness in the stomach (amplified by the fattiness of the meal) triggers a change in the muscle behavior of the gastrointestinal tract. The RPCs become GMCs, giant migrating contractions, fast, intense muscle contractions that move at about a foot every thirty seconds, traveling the length of the colon in a matter of minutes, driving digested matter before them and compacting it into what are known as boluses. One by one, these boluses are pushed—almost packed—into the rectum, where their accumulation causes feelings of distension in the stretch receptors located there. The inner sphincter relaxes of its own accord, and the pressure is then registered by the outer voluntary sphincter. We know the feeling as the urge to defecate.

The resulting feces, like the body from which it comes, is three-fourths water. The other fourth is largely bacteria, both dead and alive, along with the proteins, carbohydrates, fats, and fiber from our digested and undigested food, all of it shot through with cells and mucus from the digestive tract, small particles of the plastic that's become ubiquitous on this planet, and inorganic compounds, like the phosphates that cause feces to glow under blue light.

The bowel movement can be categorized according to the Bristol Stool Form Scale, a visual and descriptive tool introduced by doctors at the Bristol Infirmary in England in the 1990s to assist in the diagnosis and treatment of gastrointestinal disorders. The BSFS, which has since been translated into several languages, as well as modified to assist in the assessment of the stools of infants and children, divides human excrement into seven types, each corresponding roughly to transit time through the colon and representing the diagnostic gamut from severe constipation (Type 1: separate, hard, nutlike lumps) to severe diarrhea (Type 7: watery, no solid pieces).

Despite Yunmen's famous pronouncement that the Buddha is a shit-stick, there is no such thing as a perfect human stool, but Types 3 (a sausage with cracks) and 4 (a smooth sausage or snake) typify a kind of excrement that is generally characteristic of healthy human digestion. These indicate a stool that has spent enough time in the colon to allow for the breakdown and extraction of nutrients as well as the absorption of water, but that hasn't dallied there so long that it begins to dehydrate and harden. Estimates place the ideal transit time between twelve and forty-eight hours. Passed between one and three times a day, Type 3 and 4 stools, the BSFS gold standards, can be evacuated without great effort but simultaneously with control. Generally, they are drawn out and pinched at the distal end, as if sphincter and stool lingered in parting.

It doesn't seem like it should be that difficult to attain this sort of stool, and yet, as the widespread use of the BSFS makes clear, it is. People battle both diarrhea and constipation. They lose weight, develop hemorrhoids, suffer from pinched nerves. While there's no magic bullet for achieving a Type 3 stool, there are some things a person can do. For those at the lower end of the BSFS, there are a number of methods of goading the bowels into activity, from capsaicin, the irritant found in chili peppers, to senna tea, to the small sorbitol bombs known as prunes, though one must exercise some caution with sorbitol; in a case study, a woman's yearlong struggle with intractable diarrhea was ultimately traced back to her pack-a-day sugar-free chewing gum habit. Those at the looser end of the scale can ingest bismuth subsalicylate, better known by the brand name Pepto-Bismol, a pink sludge, the precise action of which is still not fully understood.

Fiber, of course, is the great champion of regularity, and you can dose yourself with it in any number of ways, from the wormlike curlicues of Fiber One cereal to gel caps filled with dried ground psyllium, an especially fibrous Eurasian plantain. You can also eat a lot of fruits and vegetables. But not all fiber is created equal, and depending on the kind of fiber, it can have dramatically different effects. Fiber occupies a spectrum, with soluble fibers at one end and insoluble fibers at the other. Insoluble fiber—think of the tough skin on a corn kernel—resists digestion and fermentation. It keeps the stool bulky, and by abrading the walls of the digestive tract, it stimulates the release of water and mucus (plastic cut to match the size and shape of insoluble wheat bran has been shown to have the same effect as wheat bran itself). Certain soluble fibers such as those found in oatmeal, on the other hand, absorb water, giving the stool a gel-like consistency that resists dehydration in the colon.

It's hard to say why defecation often vexes us to such a degree, but it's revealing that one of the most common causes of gastrointestinal dysfunction is simply not being home. This is part of the reason travelers experience constipation and children starting school or living through divorce often struggle with GI dysfunction. Being home is a potent stimulus, so much so that just putting the key in the front door lock can be enough to loosen the sphincter. Maybe this association of home and defecation is the reason the oldest extant feces in the genus *Homo*, a fifty-thousand-year-old Neanderthal coprolite, was discovered on an ancient hearth in Spain, deposited there just before the defecator's departure, archaeologists believe, as a way of preserving the camp for a return that was never made.

This Too Shall Pass

We should not fear death. That's what Socrates says. To do so would be to presume to know what lies beyond this life.

Philosophy, he maintains, is how one prepares to die, a not entirely surprising perspective for a soldier who survived three years of plagues, frostbite, and cannibalism in the siege of Potidaea. But it seems to me Alexander the Great internalized this injunction too fully, rushing headlong into the melee with no provisions for a successor, so that when he died in the palace of Nebuchadnezzar II in Babylon—his intestines perforated, some believe, by typhoid—the world was plunged into a tumultuous contest of generals and heirs from which it has not yet recovered.

The pharaohs of Egypt opted for a diametrically opposite approach, dedicating their lives to the construction of their sepulchers and crowding their tombs not only with their beloved belongings—ivory combs, scepters, entire boats, carefully fashioned furniture—but also with those people whom they felt in death, presumably, they could not survive without. Perhaps only the burial chambers of ancient China rival those of Egypt in their degree of thoroughness. In these subterranean rooms, modeled on the houses of the living, all the necessities of life were laid away, so that the dead, dressed in suits of jade and golden thread, could go on exactly as they had before, preparing food, taking the servants to task, lighting incense in the evening, regarding themselves in the mirror, and retiring when necessary to a little room set aside for the relief of the bladder and bowels (which arrangement recalls, somehow, the Borgesian approximation of eternity in "The Library of Babel" and the tiny compartments tucked between the endless hexagons "for satisfying one's physical necessities").

The question of whether one would defecate in death isn't so peculiar; both acts—dying and defecating—represent the foremost illustrations of our bodily disintegration and decay, and the historical record evidences a long entwinement of the two themes. The Talmud, for instance, envisions a hell of boiling excrement, and the eschatology of Martin Luther was summarized most succinctly in his purported dinner table comment: "I am like a ripe shit, and the world a gigantic asshole. Soon we will part." According to Thomas Edison, it was not sloth or avarice but dyspepsia that was "Satan's principal agent." One of the main duties of Tlazolteotl, the Aztec goddess of sin and forgiveness, was the devouring of human filth, and when depicted, her mouth is shown encircled by a ring of ochre. As Deuteronomy shows, early religion was preoccupied as much with the disposition of our effluvium as the disposition of our souls:

> You shall have a designated area outside the camp to which you shall go. With your utensils you shall have a trowel; when you relieve yourself outside, you shall dig a hole with it and then cover up your excrement. Because the Lord your God travels along with your camp, to save you and to hand over your enemies to you, therefore your camp must be holy, so that he may not see anything indecent among you and turn away from you.

This physical escape from the fecal had its spiritual corollary, as well. Medieval Christian theologian Giles of Rome was fond of debating whether certain parts of the body would be resurrected upon the second coming of Christ. Would the fingernails, for instance, or the genitals and intestines? And if the intestines were to rise—one must carry this line of thinking to its conclusion—would the matter inside them also be transported to heaven, and what role would this risen material play in the eternal order?

Excrement has from time to time provided auguries. One of Napoleon's generals once noted with some satisfaction in his journal that the emperor had succeeded that morning in passing a smooth movement, a sign that perhaps the day's battle would go well. Billy Wilder recalled that legendary director Ernst Lubitsch received his greatest inspiration on the toilet: "We noticed that whenever he came up with an idea, I mean a really *great* idea, it was after he came out of the can. I started to suspect he had a little ghostwriter in the bowl of the toilet there." Friedrich Nietzsche believed that after "every *creative* deed, every deed that issues from one's most authentic, inmost, nethermost regions," the body "digests less well, does not like to move." And after being treated for a deeply seated retentive urge by Carl Jung, Sabine Spielrein herself became a psychoanalyst, occupying for a time the third vertex in a murky love triangle with Jung and Sigmund Freud, in the process laying the foundation for, or perhaps even formulating, Jung's theory of the anima and Freud's of the death wish.

Often feces, either by its reticence or by its garrulousness, is the very bedfellow of death. John Wayne is said to have died with a large quantity of impacted feces in his colon, and Elvis, some believe, passed from his earthly kingdom while attempting, via the Valsalva maneuver, to move the hardened stool that is the unwavering companion of the dreamy-eyed opium-eater. Napoleon, whose constipation gave him hemorrhoids, forcing him to ride sidesaddle, died forsaken on St. Helena complaining of his insides. And if not in constipation, then illness approaches us like it did Siddhartha Gautama, from the other end of the Bristol Stool Form Scale. As was said of the ascetic Arius, death finds its figure in a violent relaxation of the bowels.

Not only does the well-being of the individual hinge upon regular evacuation, so does the well-being of civilization itself. The city too is a kind of body, and one of the foremost impediments to civilization has always been the matter of sewage disposal. In the era of the flush toilet, it's easy to overlook the rapid accumulation of feces in areas of high population density. But if you have visited a porta potty at a music festival, you know the consequences of congregating en masse in the absence of sewer systems. Excrement very literally piles up. As the plumber's adage goes, there are only two truths in this world: payday is Friday and shit runs downhill.

The Sumerians, in cities like Ur and Nippur, were some of the first to face this problem of accumulation, and in response they created some of the earliest indoor toilets. Sometimes located in small rooms beneath the stairs, these toilets were often little more than holes in the floor. Occasionally surmounted by a pair of mudbrick foot stands or a brick seat, they opened onto either large pits or short sloped pipes made of interlocking ceramic tiles. But this was hardly a sewage system. The pipes merely carried waste through the nearest wall and into the street, where it awaited the next rainstorm. The first true sewer system is generally regarded as an invention of the Minoan civilization on the island of Crete, a complex of tunnels beneath the palace at Knossos so elaborate that the man who excavated them thought he'd found the famed labyrinth of Daedalus. The ancient Greeks and Romans carried these Minoan innovations forward, creating larger and more complex sewer systems, and in the first few hundred years AD, the capital of the Han dynasty featured a highly developed system of troughs, channels, and drainage ponds for its roughly half a million inhabitants. Baghdad, the largest city on the planet from the middle of the eighth century until the middle of the tenth, derived no small part of its success from its location at a bend in the Tigris. Surrounded by canals and pierced by water conduits, it was said to contain sixty-five thousand baths, though its status as one of the world's great sewerages remains as yet hypothetical, modern archaeological excavation of the ancient City of Peace having been rendered impossible by the United States invasion in 2003.

Despite advances in hydrology and sanitation, few cities evolved with any framework in place for the disposal of feces. Instead, an ad hoc approach predominated. Polluted waterways were covered, becoming de facto sewers, and the streets themselves doubled as open-air sewers. As Jonathan Swift wrote following a spring downpour in London in 1710:

> Sweepings from butchers' stalls, dung, guts, and
>   blood,
> Drowned puppies, stinking sprats, all drenched in mud,
> Dead cats, and turnip tops, come tumbling down the
>   flood.

Even well into the nineteenth century, many cities in the United States and around the world had no sewage systems whatsoever. Instead, in cities like Baltimore and Washington, DC, there existed a class of laborers known as night soil men, whose sole occupation was digging out privies under cover of darkness and transporting their contents by wagon beyond the city limits. It wasn't until the Industrial Revolution, with the dramatic crowding of cities and the discovery of germ theory, that sewage was recognized as an existential threat to civilization. One gets a sense of the state of sewerage at the time by realizing that when Eugène Belgrand set out to build a comprehensive sewer system for Paris following the epidemics of the 1830s, his first task was simply to map the incoherent maze of troughs and tunnels by which the city then disposed of its waste.

The systems created by engineers like Belgrand continue to function today, but having been designed to use stormwater to carry waste from the city, they are now often overloaded by the heavy rains that have become more common with climate change. Following a downpour, the runoff from streets and buildings overwhelms the capacity of treatment plants, and torrents of untreated excrement surge into waterways, as happens with astonishing regularity in New York City. If you want to confront the original dilemma of the Sumerians, you need only kayak after a rainstorm on Newtown Creek. A short, shallow stream dividing Brooklyn from Queens, in a matter of hours, it is transformed into a swamp of bubbling, decaying organic waste, its depths obscured by brown murk, its surface paddied with the grayish rags of flushable wipes.

The danger, of course, lies in what goes largely unseen in such a morass. Though an excellent fertilizer, human feces often contains a catalog of pathogens reminiscent of the contents of Pandora's box: everything from parvovirus, which causes the mild temporary skin condition known as fifth disease, to hookworm eggs, and from *Salmonella typhi*, the bacterium responsible for typhoid fever, to poliovirus, which, before the advent of the polio vaccine, indiscriminately sickened and paralyzed children around the world. Even in this hazardous mixture, however, certain members of the bacterial genus *Vibrio* are notable for how well our waste suits them.

One cannot see these bacteria, but at something like 750x magnification, they become quite apparent, small tubes with icosahedral heads and long contractile tails. They exist throughout the aqueous parts of the world and can often be found in the company of plankton and mollusks. They live a carefree life, following the currents, reproducing, in difficult times encapsulating themselves in a protective biofilm; occasionally they are drunk or eaten by humans. Many do not survive the hostile acid environment of the stomach, but those that do are well adapted to life in the lower tracts, where they burrow through the mucosal lining and embed themselves in the crevices of the intestinal wall. There, after multiplying rapidly, they secrete a toxin, the specific nature of which causes the intestines to exude water in vast quantities. The rapidity of this fluid accumulation is such that the bowel cannot reabsorb it, and soon it flows out of the host in a painless grayish stream, colloquially known as rice water. This effusive liquid stool, which is sometimes both the first and final symptom of a *Vibrio* infection, carries with it on its return to the world uncountable numbers of the bacteria. The onset of the illness is so sudden, and the liquid so voluminous, that the mere containment and disposal of the effluvia is itself a monumental task, best accomplished with rubber sheets, fully enclosed protective gear for those nearby (the term "hyperinfectivity" is sometimes used to describe a certain and particularly virulent stage of the bacterium's life cycle), vast quantities of disinfectant, and a great number of buckets. This diarrhea continues at a rate of as much as a liter per hour until, within as little as a few hours, the body begins to suffer from acute dehydration. The eyes sink into the sockets, the blood thickens in the veins, the skin becomes drawn and inelastic. We call the disease cholera. If, at this moment, intravenous fluids are not administered rapidly, death often waits close at hand. *Vibrio cholerae* does not require the survival of the host. The human body, to us the very central purpose of the universe, is to these bacteria nothing more than a dark and incubatory lull in the course of their own watery peregrinations.

It isn't clear how long cholera has accompanied human civilization, but for the past two hundred years it has been a near constant companion, producing in that time at least seven separate pandemics and countless epidemics. In the nineteenth century, it killed millions, stalking the battlefields of the American Civil War and the pilgrims' paths home from Mecca. On the island of Java alone, it claimed a hundred thousand. It ravaged London with such regularity that it helped usher in the era of germ theory and the branch of science known as epidemiology. Even so, since 1961, by the WHO's accounting, the world has been trapped in the seventh cholera pandemic, this one largely caused by the El Tor strain. The disease is kept at bay by sewer mains and oral rehydration salts, but it is not gone.

Epidemics of cholera continue, often following like a second army in the wake of war and natural disaster, when the great infrastructures of plumbing and medicine begin to fail, as we have seen recently in Haiti and Zimbabwe and as we are told is now occurring in certain parts of Yemen, where the greatest epidemic of cholera in human history is progressing with mathematical efficiency, encompassing something on the order of a million human beings. Faced with these numbers (four thousand infected each day, of whom half are children), it can be difficult to account for the relatively minor public reaction to a disaster that is, after all, in large part a human-made tragedy, the result of a war led by Saudi Arabia and supported by the governments, and implicitly the citizens, of the United Kingdom and the United States, though perhaps it is not entirely surprising, considering how often we take for granted the vast networks of excrement that make our modern lives possible. *Inter urinas et faesces nascimur*, as Freud tells us, quoting a church father. *Between piss and shit we are born*, to which we might easily add, *et morimur, and we die*. And it strikes me that the old theologians, whose heavens could not quite compass excrement, might applaud those who, by dying with their bowels empty, so simplify a long-standing paradox in the concept of redemption.

X

Vomit

As a dog returns to his vomit, so a fool returns to his folly.

Proverbs 26:11

Terms

n. barf, puke, throw-up, spit-up, mess, chunks, lunch, cookies

v. barf, yarf, spew, spit up, burp up (mostly in infants), throw up, bring up (with noun; e.g., lunch), chuck up, upchuck, chunder, yak, ralph, regurg, blow chunks, boot, bow down before the porcelain goddess, drive the porcelain bus, Technicolor yawn, bushusuru (Japanese, translating literally as "do a Bush," after former president George H. W. Bush inadvertently vomited on then-prime minister Kiichi Miyazawa), lose one's lunch, toss one's cookies, revisit one's dinner, puke (from Shakespeare's *As You Like It*, "All the world's a stage / And all the men and women merely players. . . . At first the infant, / mewling and puking in the nurse's arms"), heave, hork, retch, be sick

Etching of a man vomiting into a bowl as his companion lifts his wig and steadies the bowl, *Human Passions Delineated*, T. Sanders and John Collier, 1773

Biological Prologue: Birth of an Emesis

Ordinarily, it takes about three hours for the stomach to complete the initial breakdown of a meal and pass its liquefied products through the pyloric sphincter into the small intestine. Every so often, however, the matter never makes it through the sphincter. The condition is fairly easy to diagnose even without a stethoscope. Put a hand on your abdomen and rock your body side to side. If you can feel or hear a sloshing in your stomach—what's called a succession splash—it's a sign that your most recent meal has been detained in its digestive journey. Possibly, the pyloric sphincter is blocked by a tumor or a foreign object. Or, in the continual chatter that occurs between the digestive tract and brain—the gut-brain axis—it's possible your body has concluded the matter in the stomach should proceed no farther. In either case, whether you feel ill yet or not, the succession splash is a decent sign that vomit's in your future.

When you're going to vomit, you may experience nausea. The heart will begin to beat rapidly, and waves of heat will pass through the body, causing sweat to bead up from glands in the skin. The mouth will begin to water, as well, and this probably is the best sign that the moment of emesis is at hand. Vomiting does not come immediately, though. It is preceded by retching, a setting of the emetic stage. The abdominal muscles and the diaphragm contract, causing elevated pressure in the stomach and reduced pressure in the thorax, basically raising matter so that it's well-positioned for ejection. Then the diaphragm relaxes, the intercostals and abdominal muscles contract dramatically, and the material is brought up through the esophagus, the throat, the mouth and returned to the world from whence it came.

The vomit itself is largely composed of your last meal, along with mucus, hydrochloric acid, lipase, and pepsin from the stomach. It may also contain a great deal of the toxin, virus, or bacteria responsible for the reaction. If a vessel in the stomach ruptures, there may be blood. If retrograde contractions in the digestive tract have brought chyme back into the stomach from the small intestines, there may be bile. Occasionally, in exceedingly rare cases, an intestinal obstruction can cause feculent emesis, the vomiting of feces, easily diagnosable by the smell of the breath.

As for the causes of emesis, they are as varied as they are vast. Pain and stress can cause vomiting, as can arsenic. Cancer treatments often result in emesis, as does general anesthesia. A person can vomit because of underlying disease, whether ulcers or xanthinuria, or they might vomit because they swallowed a quarter. Vomiting can be a positive sign, as in the morning sickness of the first trimester, when mild emesis indicates the shift in hormones necessary for a healthy pregnancy. Or it can indicate the rampant reproduction of a virus or bacteria within the digestive tract, perhaps most notably the norovirus, which can spread via aerosolized vomit and is often known as the cruise ship virus due to its propensity for ravaging entire ships, as it did in 2019, when it turned the *Oasis of the Seas* into a vomitous floating Caribbean quarantine. People get sick due to the sensation of motion, whether aboard trains, cars, airplanes, trampolines, or at 3D movies, though this form of sickness is most often associated with ships. (The word *nausea* itself stems from the Greek *naus,* or ship, which also gives us words like *nautical* and *navigation*.) Not too infrequently, people vomit because they drank too much champagne.

Most interesting, though, in terms of the causes of emesis, is the relationship between nausea and vomiting. Nausea often precedes vomiting, and while we often think of the former as being one step in the progression of symptoms that ultimately ends in the latter, they are actually the result of wholly independent neural processes, which is why a person can vomit without feeling sick and feel sick without vomiting. Theoretically, these two reactions act as complementary methods of self-preservation. Vomiting removes noxious material from the body and, in the process, encodes the accompanying sensations into memory, the result being that certain sights, sounds, and tastes produce in us so strong a bodily aversion that we flee the stimuli. But nausea itself is powerful enough that some species, rodents for instance, wholly lack the ability to vomit. Instead, much like toddlers, they are extremely phobic of new tastes. Presented with a novel food source, rats will nibble and wait. If the food induces no nausea, they may return to it. If nausea occurs, they engage in pica, diluting the stomach's contents by eating nonnutritive substances like clay, and then they never eat the food again.

So powerful is nausea, in fact, that merely driving by the hospital where they receive treatments can cause chemotherapy patients to vomit. In one notable case report, an eleven-year-old girl was sickened with pneumonia in December 1986 and spent that Christmas in the hospital vomiting. The following December 23 found her hospitalized once more. Again, Christmas arrived with a torrent of emesis. "Having spent two successive Christmases in hospital," the case report reads, "the patient hoped to be well the next year." But it was not to be. "She was woken by nausea in the early hours of Christmas Day 1988." This attack, however, was precipitated by no illness. It was Christmas itself, with its potent cultural stimuli of carols, cookies, lights, and spruce, that made her sick.

I Did the Hardest Thing in My Life in a Porta John in Orlando

On a spiritual map, Soul Quest Ayahuasca Church of Mother Earth sits firmly at the nexus of our world and several less-burdened dimensions, but on this plane, it's located about twenty minutes east of downtown Orlando, on a long skinny parcel of live oaks and bamboo, between a martial arts academy and an auto body shop. The conventional way to arrive is by automobile, usually a Lyft from the airport, though occasionally people make the long haul down from Georgia or Alabama in their SUV and park behind the horse gate in the lot out back.

Despite the church's lack of attachment to the usual terrestrial expectations, it is not a fly-by-night sort of affair. There's a tent in the back where Sean, a beefy medic with a silver buzz cut, takes temps and blood pressures and verifies that new arrivals for the weekend retreat have had a negative COVID test within the last seventy-two hours. And someone in the organization knows their way around Microsoft Word: in the front office, there are printed name tags with the church logo and designations as to the wearer's role and lodging details. While the church does not charge for its holy medicine—a bitter emetic that transports congregants into a state of hallucinatory barfing—they do collect a donation for the two-day retreat, about $900, to cover the cost of accommodations (a four-inch foam mattress on the floor), meals (intermittent celery, sweet potato chips, rice, broth), and "the work of accompaniment and supervision." To facilitate the transfer of this lucre, the church is connected to the usual payment channels. They accept Visa and Mastercard, even if they occasionally mix up the files and inadvertently try to charge you for the extra ayahuasca ceremony on Saturday morning.

I spent some time in mid-September as a member of this church. Along with eighty-five or so other souls, I swiped my credit card, accepted the sacrament, and embarked on a journey. I have no doubt that some of my fellow travelers found a measure of relief in the arms of ayahuasca on the grounds of that church. I only know that my single strongest conviction as I crept out under the locked horse gate was that I would never, not in a million celestial lifetimes, set foot again on that property. I am someone who failed to transcend. My ego was not burned from my bones by the juice of the vine, I met no earthen goddesses, vomited no endlessly coiled snake. I write this because I did not make it to the other side and some did, and they're no longer around to give an accounting.

Even now, hurtling north on the Amtrak Silver Meteor,

separated from the retreat by little more than a handful of hours and sixty miles of steel, I cannot recall the exact shifting in my perspective that occurred over the course of the weekend or how precisely I found myself, ant-bitten and half-conscious, being evacuated by stretcher from a grove of oaks out back of the compound. When exactly did Rayn start to seize? And was it before or after my hands grew two sizes? Some answers only get harder to grasp. Someone made the decision to join that church, and someone writes these words now, but I'm unclear what precisely separates those two people. This uncertainty gives to the process of composition a disconcerting and unhelmed quality. I can say only one thing with any surety—and this I mention in the hope of sufficiently inculpating myself in what turned out to be the single worst decision of my entire life: in retrospect, there were a great many hints.

I CAME UP WITH THE IDEA OF DRINKING AYAHUASCA on a whim. I don't know where I'd first heard of the stuff. I'd been researching and writing about vomit, and one morning, as I woke up, I seemed to remember a mention in some academic article of a syncretic Amazonian religion whose sacrament was both a physical and a spiritual emetic.

The whole scheme was always going to be a hard sell with my wife. In certain respects (licensed in the state of Iowa to instruct yoga, occasionally listens to Phish of her own volition), she's got way more hippie vibes than I do, but she's also got this uncanny taste for equilibrium. Given the choice between a weekend of vomitous psychic exploration and buying a couch, she'll go couch every time. Whatever the opposite of stoked is in the lexicon of psychedelia, that was Erin's general feeling about this whole endeavor. We were in the living room when I pitched her on the idea, in the place where the couch would have been if we had a couch. Erin has a knack for words when she's trying to make a point. She looked at me and summed up all her reservations in a single question. "What are you trying to accomplish?"

It's a solid question. I mean, on the surface, I'd been asked to write about vomit, and I'd found a way to do it that justified spending close to nine hundred tax-deductible dollars on drugs (though again, technically, the drugs were free).

Many people find their way to ayahuasca because they hope to overcome some physical or mental trauma, but my health and emotions were generally in pretty good shape. Not much traumatic experience under my belt besides the usual collection: a crashed car, a couple of buried dogs, and a handful of human funerals. I was in good health too, with the exception, entirely writing induced, of steadily worsening vision in my left eye. Other than some low-level daily stress, largely related to the pressures of trying to make a living while managing the emotive whims of a toddler in an era of social turmoil and chaotic global weather, I was pretty satisfied with my existence. I was, as Erin, buttering me up, told me later that week, "the most psychologically resilient person" she knew. "Sometimes annoyingly so."

But I wasn't interested in healing. I didn't want to commune with the earth mother. I was interested in throwing up.

Ayahuasca is a strange drug. It's not like popping a Xanax or an Adderall. It's not like eating mushrooms or knocking back a sixer of beer. There's nothing recreational about it. Ayahuasca's whole reputation was that it made you puke. The drug was reputed to induce not merely intense physical vomiting but also a kind of puking beyond puking, a sensation of purging from the soul itself, and having set out to write about vomit, I thought I should experience the mother of all puking. I didn't expect to be healed or remade or become a better person. I just wanted to see what this brain of mine would vomit up.

It was sentiments like these, more delicately worded, that I expressed to my wife little by little over several days in July, during that period when the money was still refundable, and little by little she warmed to the idea, or rather resigned herself to it, inaction being a strength we both share.

"So it's three days."

"Yep."

"And the drug is safe."

"In the traditional sense."

A deep, couchless sigh.

"It's only Orlando," I said.

"Orlando?! You can't just do your drugs here in Baltimore like everybody else? You have to go to Orlando?"

WHILE THE USE OF PSYCHOACTIVE SUBSTANCES HAS historically been afforded few legal protections in the United States, freedom of religion is one of those liberties vouchsafed by James Madison, Thomas Jefferson, Thomas Paine, et al. in the earliest days of the nation. As written in Jefferson's typically wordy and idiosyncratically punctuated Virginia Statute, religious freedom is necessary for many reasons, the ultimate of which is simply this:

> That Truth is great, and will prevail if left to herself, that she is the proper and sufficient antagonist to error, and has nothing to fear from the conflict, unless by human interposition disarmed of her natural weapons free argument and debate, errors ceasing to be dangerous when it is permitted freely to contradict them.

So central is this freedom to our legal system that, should members of the government wish to in any way abridge it, they must provide evidence of a compelling interest, basically some reason that leaves them with no other choice.

It was precisely this compelling interest that the government's lawyers were unable to demonstrate to the Supreme Court in 2005, in *Gonzalez v. O Centro Espírita Beneficente União do Vegetal*. The UdV was a religious society, founded in Brazil in 1961. Its members practiced a form of religion that combined elements of Christianity with the indigenous Amazonian ritual of drinking ayahuasca, and at the time, it had chapters in a number of countries, including one with offices in New Mexico. In 1999, federal agents raided the church's offices in Santa Fe, seizing close to thirty gallons of ayahuasca tea that the members of the church used for their ceremonies. The action would quite likely have gone uncontested had the church's *mestre* not been Jeffrey Bronfman, one of the heirs to the Seagram's fortune. Bronfman and the UdV took the government to court in a suit that spanned the end of the Clinton presidency, the 2000 presidential election, the Florida recount, September 11, 2001, and the beginning of the wars in Afghanistan and Iraq, steadily working their way up the layers of the judiciary until finally the Supreme Court ruled unanimously in their favor. In his opinion, Chief Justice John Roberts noted that exceptions to the controlled substances act had long been made for the use of peyote by Native Americans, writing with no small amount of sass, "The Government's argument echoes the classic rejoinder of bureaucrats throughout history: If I make an exception for you, I'll have to make one for everybody, so no exceptions."

While bolstering the cause of religious freedom generally and the rights of native religions specifically, the most practical and immediate result of that 2005 decision was that drinking ayahuasca in the United States was no longer some shady backroom affair; it had the legal imprimatur of the highest court in the land. There was no mad rush to do so—as I mentioned, the most notable feature of ayahuasca is its propensity to make users vomit—but in the years following, a few churches quietly opened their doors. These days, there are a number of places in this country where a person can go drink ayahuasca. Should you so desire, you can probably drink ayahuasca somewhere this weekend. You can puke to your heart's content, relive past lives, swim the length of the Styx, and meet the Buddha in 4D. You can do it all, provided you're a member of a church taking part in a sanctioned religious ritual, which is how I came to join Soul Quest Ayahuasca Church of Mother Earth in Orlando.

It was a simple process, really. By signing some forms online, I accepted the SQACME's spiritual and religious principles, becoming, with a click of the mouse, "part of Mother Earth, an indivisible, living community of interrelated and interdependent beings with a common destiny." I accepted that this destiny was being "crushed by the capitalist system and all forms of depredation, exploitation, abuse and contamination," and I took as my holy book the *Ayahuasca Manifesto*, an anonymously authored avocado-green scriptural pdf and ceremonial user manual written from the perspective of the spirit of ayahuasca:

> I operate from a vibration superior to the spirits who compose me. . . . I am the mixture in its natural state, crude and basic, bio-electrically loaded, without industrial processing. . . . I come to assist certain little known processes of the Human genetic code, the DNA. Specific subatomic codes that are inaccessible from the third dimension are becoming active.

And although I have never been quite clear what constitutes human consciousness, I consented to the *Manifesto*'s insistence that there is an overwhelming need for consciousness expansion, a process that "consists in living its cosmic existence beyond its three-dimensional physical limitation, with a very subtle vehicle capable of moving consciously at will, even while in the company of others vibrating in the same channel." I also consented that, as a tithe to the church, 10 percent of my retreat cost would be added to my booking "atomically."

The difficulty, of course, lies not in accepting one's faith but in living it, and as the date of the retreat approached, I struggled to abide by the checklist emailed over by SQACME. It was encouraged that I keep a humble heart, clear intentions, an open mind, but more than this, the preparations to enter a cosmic existence beyond three dimensions mainly involved, as far as I could tell, not eating a bunch of stuff. No salt, no sugar, no artificial sweeteners, no red meat, pork, animal fats, fermented foods, dairy or caffeine, no eels, no carbonated drinks, no spicy food or processed food, no fried food or dried fruit, and no pickles, no herring, no anchovies, no old cheese, no too-ripe fruit. Also, no drugs, alcohol, sex, or masturbation.

Some of these tenets were observed with ease, eels not being easy to come by at my grocery store in Baltimore, but others proved physically and philosophically challenging. It was the very end of summer at the time, those few weeks of the year in the Mid-Atlantic when the earth offers up its vegetal booty to every passerby, and everywhere you go there is a tomato hanging on the vine threatening to burst if you look at it too long. There was a pile of white peaches on the kitchen counter that week. Their flesh seemed to yield to the action of the thumb with an almost wanton delight. How ripe was too ripe? What was ripe, after all, but the imposition of a human order on the natural life of the fruit? The juice of a peach spilling down the throat, the juice of a peach spilling over the lips, the juice of a peach running down one's arm to the elbow, a tongue in pursuit, wasn't this really just the sort of thing the earth mother had in mind? And also what exactly qualified as sex? The SQACME email had encouraged us to "take some time to sit quietly and ask yourself some questions," but I strongly suspected these were the wrong questions to be asking.

Now, faith unchallenged is hardly faith at all, but for those of us experiencing such trepidations in the run-up to the retreat, the church helpfully provided weekly group pep talks on Zoom. These talks, unfortunately, coincided exactly with that period of the afternoon when my daughter, having returned from pre-K in a state of hypoglycemic mania, prowled the homestead, rooting out and systematically dismantling all vestiges of calm. For much of the roughly hour-long meeting, I sat at my desk in a rotating chair while my daughter spun me in circles, making my experience less reassuring than nauseatingly strobe-like. Every two seconds, a muscular man with a curtain of beard appeared in my line of sight and spoke some snippet of wisdom about the modus operandi of the vine before disappearing in a blur.

"Mother's going to teach you . . . the difference between your ego and your heart space . . . surrender . . . if you don't surrender, she'll keep coming on stronger . . . she loves you . . . she's heart-opening medicine . . . she's truth . . . if anything is scary . . . that's her shining a flashlight on what's already been inside you . . ."

He reiterated the importance of abstinence. Not only were we to avoid sex and masturbation; we were to stay away from intimacy generally. In fact, any exchange of bodily fluids was inadvisable. "We're cleaning out the vessel," he said. "There's chi energy in that aura. Be cognizant of what energies you're bringing into your field."

My daughter stopped suddenly, as if for the first time noticing the screen. "What's a vessel?"

Let's be up front here. This whole situation was ridiculous. You've got a bunch of horseback-riding late-Enlightenment deists with a suspicion of theological rigmarole and sectarian tribalism. They enshrine religious freedom in the Constitution, believing that democracy can't survive without a love of reason and the pursuit of truth, and a couple of centuries later, I'm preparing to drink some deeply psychoactive Amazonian swill legally validated by the heir to a Canadian Prohibition fortune and listening as a too-buff man on Zoom tells me cunnilingus will disturb the flow of Chinese Buddhist energies in a colonial British-Indian energy field and I won't be able to surrender to the earth mother's flashlight. That's a hefty dose of baloney. This may be related to being intermittently spun in a desk chair for close to an hour, but as I listened to that Zoom session, I began to succumb to a kind of anticipatory nausea at the very thought of the upcoming retreat. It seemed to me we had somehow managed to protect the very nonsense Thomas Paine so despised while allowing the truth to disperse like so much chaff.

It had been a beautiful image, the Jeffersonian ideal of truth cutting error down with a fell and awful strike, but it was hard to believe truth a sufficient antagonist to error in the late summer of 2021, when no small number of human beings were poisoning themselves with an equine de-wormer.

THERE WAS NOTHING OBVIOUSLY REMARKABLE ABOUT the waiting room on the day I arrived at Soul Quest Ayahuasca Church of Mother Earth: intense jaguars staring out from several tapestries, a smiling Buddha, some ailing houseplants, a ceiling fan slowly wobbling itself loose from the rafters. One table had name tags, another a tray of avocado lettuce wraps. A pair of apparently normal women were working the desk, handling phones, emails, and the steady stream of newly arrived participants. On the wall by the door was one of those framed sayings that you can buy at any store where superfluous décor is sold. "Life is not measured by the breaths you take but by the moments that take your breath away."

There were maybe twelve chairs around the perimeter, each filled with someone who'd come for the retreat. We were waiting to be called into a back room and cleared by the medical staff, and for a long while, we sat in silence. With the exception of a woman on a bench by herself, who happily invited anyone in earshot into a conversation about her love handles, her sciatica, her herniated disc, her husband's keto farts, and her son who'd been taken away from her by social services, the talk was intermittent, guarded, died down almost as quickly as it began. We listened to the women at the desk process the new arrivals. They confirmed the name of each, collected payment for any outstanding charges, and asked whether the person wanted to take part in the additional ceremony on Saturday morning. The only interruption of the routine was the occasional opening of the front doors, and the entrance of one of the church's white-clad volunteers.

"And did you want to add the morning ceremony?"

A trim young man walked in wearing white slacks, a white muscle shirt, and a military-issue backpack.

"It takes place under the trees out back."

A woman in white skirts with hair bleached white who walked only on the balls of her feet.

"A lot of people like it. You get to be out there in the daytime, close to nature."

A fit Euro-couple in glossy white athleisure-wear, with the smooth plasticine smiles of Botox clinic habitués.

I can't remember how we all actually started talking. Maybe it was because at a certain point, silence is harder than speech. Or maybe it was Samvith, who could never seem to keep his mouth shut.

They'd come from all over, it turned out. There was Catherine, a prim pale Georgian who audited airplane inspection reports for Delta, and Terrence, a soft-spoken young Black man and self-proclaimed "country boy from Alabama," who now lived up around DC. There was Samvith, the jittery trucker who'd escaped the slums of Ahmedabad and gotten a woman pregnant in Oklahoma, where he now lived, and Elsie, with her blond highlights and strong white teeth, seeming to have just stepped out of a *Southern Living* spread, who held her abdomen in her two arms when she spoke of some vague emotional trauma. And then there was Phil, the hulking vet with the Mark Twain drawl and the gleaming bald head, who smiled very sweetly and whose hand trembled wildly when he reached out to shake my own.

"I have PTSD," he said softly. "And I have this tremor. I wanted my wife to come, but she wasn't interested."

"You came here before?" Samvith said to Terrence.

"It's my third time," Terrence said, one hand wrapped around the thin wrist of the other. "I came down the first time 'cause I found out my dad wasn't my real biological dad, and then two days later my brother died."

"It was good?"

"Yeah, it helped. It helped a lot."

"And why are you coming back?"

"I guess I got more to figure out?"

Samvith turned to another returning participant, a plumber named Hank in a tie-dye shirt with his hair pulled back in a ponytail.

"Do you really vomit?"

"Like, man," Hank said. "Like I was puking like crazy, I thought it was like for hours, all *kinds* of things, snakes and . . . and I don't even know what, and I thought I was going to need another bucket, but when I looked down it was like totally empty."

"It's not puking like regular puking," said Terrence quietly. "It's puking from your soul. And listen," he said, lowering his voice even further. "People shit themselves."

"Nooooo," said Samvith.

"For real," Terrence said, and he lowered his voice another register. "I almost shit my pants."

Samvith giggled and bounced in his seat. Then he leaned over to me. "Are you here to do drugs?"

He hardly waited for my answer. "I'm here to do drugs," he said. "Joe Rogan? You know the podcast? I heard about it on there. I think I will have one of these when I am finished. What is it?" He traced the shape of a halo above his head.

Terrence keened away. "I think you might just sprout some big old devil horns before this is all done."

They called my name then, and I headed into a back room for a medical interview that proved to be almost wholly perfunctory.

"How are you feeling?" the doctor said.

"Terrified."

"Oooh," she said. "Terrified's not what we want. How about nervous?"

"Sure." I nodded. "I'm nervous."

"Good." She strapped a yellow band around my wrist and sent me back out.

When I returned, someone was sitting in my seat, two too-skinny legs jutting out from a baggy hoodie, body slumped over a phone, bitten teal nails tracing letters on the screen. As I sat down beside her to gather my things, she tilted up to me a face so obscured by piercings and tattoos that it was almost impossible to see the person at all.

"Are you nervous?" I said. "I'm nervous."

"I'm so nervous," she said.

"Why'd you come?"

"My mom's friend. She's done this a bunch, and she thought it could help me. My mom's tried everything to help me. She was going to get me brain surgery. This seemed less scary than that."

Her name was Rayn.

We watched a new arrival wavering about whether or not to do the extra morning ceremony on Saturday.

"A lot of people feel like they want that extra ceremony to get the full effect," one of the women at the desk was saying, but before she could go on, an old man came into the room, the head volunteer, we later learned. He was brown and curled up as the tongue of a leather shoe, and he cut off the conversation.

"*Do* it," he said with a thick Brooklyn accent, pointing a crooked finger at the man. "Listen to me. Do the morning ceremony. You came all this way for a reason. Do it." Then he turned his finger at the rest of us. "I'm talking to everyone. Do. It. You will not regret it."

As he went to leave, he looked down at me. "Have we talked? Come with me."

He took me into a back room where a pantsless man in a corner was trying to put on pants, and he set two chairs facing each other. "Sit," he said. "Don't pay any attention to him. He's putting on pants. Why are you here?"

"I don't know," I said.

"You don't know."

"I'm curious," I said.

"You're curious."

He looked at me a long time. It was hard to tell if he was angry or just from New York. He was staring at my forehead, and then he pointed at it.

"Right there, that's your third eye. We call it the pineal gland, but it's your third eye. Mother's going to open that, and she's going to show you things."

Another long pause.

"Listen," he said. "You're not curious. You were called."

He continued to search my face, forehead tilted toward me with a piercing pineal gaze.

"I've been doing this a while," he said. "I don't think it's going to be too hard for you."

I COULD SPEND A GREAT DEAL OF TIME DESCRIBING the various lodges at SQACME, the hedge of invasive bamboo and the main yurt-like maloca with its central pit of Astroturf, its window AC units, and its altar scattered with ten kinds of colonialist-ethnobotanist nonsense, but let me leave it at this. The churches of Italy are made of marble and colored glass. The Jews in Jerusalem have their wall, the Muslims in Mecca their Kaaba with its oiled stone and golden rainspout. The disciples of the Buddha have their tree of awakening, well-fenced. The main structural components of Soul Quest are two-by-fours and spray foam insulation. It doesn't matter much where you are at Soul Quest, if you are indoors, chances are you are staring into a blobby field of hardened isocyanates the color of whipped egg yolk.

Having said that, though, I also want to say that I felt a real joy and warmth in the place that first afternoon. The process of finding your name tag, your lodge, the bathroom; the dropping of a bag beside an unfamiliar bed; the work of attempting to find common ground with utter strangers ("You're in Dragonfly Lodge? Me, too!"); the smell of decayed wood; the wandering around; the waiting for hot water that never comes. All of these gave to the day a distinctly summer campy atmosphere.

There was a camp-like hierarchy to the place too, which emerged as the afternoon wore on. At the bottom were folks like myself, first-timers. We wandered around aimlessly, sat down in a chair, wrote two sentences in a notebook, and jumped back up to wander some more. Above us were those who had done retreats previously. They were calmer. They moved around less, sat in a corner of the courtyard dispensing wisdom. "After a couple of hours, they're going to ring this gong and ask you if you want a booster," one man told me. A booster was a second dose. "I took a booster my first ceremony. That was my last booster."

Above these were the volunteers. These were former participants who had agreed to help with a weekend's chores. Dressed in white, they crisscrossed the compound briskly, carrying towering stacks of buckets or armfuls of firewood. They would be accompanying our bodies after we drank the brew, acting not as spiritual guides but physical ones, escorting us to the bathroom, distributing blankets, bringing water to those who needed water. In exchange, they would receive $300 off their next retreat.

Above the volunteers, there were the facilitators. One by one, they began to appear, each in their own manner. One slipped quietly in a side door, another strolled into the courtyard, pausing to hug each person he recognized and to look deeply into their faces. These were the people directly associated with the church. The *Manifesto* capped the number of participants in any single ceremony at twenty-five, and so each of these facilitators would oversee a portion of the larger group during the evening's ceremony. They were not, according to the *Manifesto*, shamans. "Modernism has arrived," the pdf said, "making original indigenous shamanism scarce," but "it is not necessary to lament the gradual disappearance of genuine ancient shamanism as you known [sic] it." Facilitators were the inevitable evolution of Mother Ayahuasca's spread through modern civilization. They were to be our keepers on our psychic journeys. Only they could "create a protected energy space." They were the top of the ladder, unless you counted Mother.

But even this weird hierarchy didn't bug me the way it normally would because the thing was, everybody was really nice. Although I couldn't help doing the mental math on the organizational budget (at ~$1,000/person, that would be $80K/weekend, grossing say $4M/year), there was a real charm to the people milling around. Sure, they hugged a little more than the average population sample. They looked more closely at my face than most people do and nodded a little more intently than I was accustomed to, but what's wrong with hugging and looking and listening? Flip-flops were de rigueur, and there was just something sort of fun about the fact that we were all about to take part in an activity that anyone in their right mind could see was deeply questionable. The only thing that caught me off-guard was that, while the crowd hailed from every segment of society, there was a slight overrepresentation of muscle-bound dudes. Not just fit, but the sort of jacked physique that takes both some serious gym time and also maybe a slight body-image disorder to achieve. I didn't think about it too hard. I knew that ayahuasca was sometimes used to combat PTSD, and I figured maybe that explained it. And anyway, even these dudes had about them a gentleness. I saw two of them do one of those hugs where you hug then draw away, holding the other person by the upper arms, and talk for a while with your faces only about twelve inches apart.

Part of my warm fuzzy feelings probably had to do with my own internal mental state. Though I didn't think too hard about it then, I was already experiencing the sense of abstraction from the world that a drastic reduction in calories inevitably produces. The spinning Zoom call, the eighteen hours rocking side to side on the train down from Baltimore, and the strongly laxative quality of Amtrak's soy meat enchiladas had also left me with a mild but persistent feeling of vertigo. I felt light, hollowed out even, and I moved slowly. In the outdoor bathroom before the first ceremony, straightening after washing my face, I had such a strong sensation of the floor shifting beneath my feet that I immediately stepped outside and checked to make sure the building wasn't on a raft.

This was abnormal behavior. I am not someone who gets dizzy. People do not ordinarily check to see whether a building is sitting on the ground. I mention it now because I did not quite register its abnormality at the time, and I think my lack of awareness of my own functional degradation goes a long way toward explaining just how zenned out we all were. Late in the afternoon, as the group slowly gathered in the central yurt for a mandatory orientation, I saw a man in a numberless football jersey peering in the windows, unable to enter the sacred space, having pulled with all his strength and determination on a set of doors that only opened inward.

AROUND DUSK, WE ALL GATHERED IN THE COURT-yard. A hedge of bamboo swayed above us. Someone lit a fire in the firepit. The volunteers started waving bundles of smoking sage around our bodies and spritzing our hands with Florida water. Our purifying complete, we were instructed to take a handful of uncooked rice from a large bowl, filling the grains with our intentions, and to throw this rice into the flames. Then they gave us each a little white cup with a slosh of thick brown sludge inside.

We'd all grown accustomed to the terminology by then. The phrases came to mind naturally. Staring down into these cups we knew what we were looking at was not a drug. This was the *medicine*. The medicine would deliver us to Mother Ayahuasca. Tonight, we'd been told, we would meet her. Tomorrow morning, we would date her. (It was presumed by this time that we had all signed up for the additional ceremony; I had.) Tomorrow night, the two of us would marry. It was not vomiting that would chaperone this courtship. It was purging. This was an important distinction, we now understood, for one thing because purging connotated a more spiritual experience than vomiting. More pressingly, though, we'd learned little by little over the afternoon, it was called purging because many people did not vomit at all. Instead, they had explosive diarrhea. "We have a little saying in this business," one of the facilitators told the group, his last word of advice before releasing us to the ceremony. "Never trust an ayahuasca fart."

The *Manifesto* says those who most need Mother Ayahuasca shall find her by their vibratory attunement, and as we all milled about the courtyard, staring down from time to time into the viscous liquid in our cups, there *was* a certain kind of universal vibration, though it would be perhaps better described as collective agitation. Those who had been calm were now nervous. Those who had been gregarious became quiet. Terrence, delicate Terrence, had ceased to see other people. He passed through the crowd like a Shakespearean ghost. One old woman stood alone, looking lonely. I went to her. We talked. I don't remember what we said. I moved on. The dominant motif was turmoil. Cups in hand, we returned quietly to our lodges and our mattresses, preoccupied by what lay ahead.

And there was reason to be concerned. While tryptamines like those in ayahuasca are generally pretty safe, there's no real guarantee that *you'll* be safe while you're on them. There was the volunteer who told me in passing that a spaceship had tried to abduct her during her last trip. "Just surrender," she recommended. "Otherwise, you'll only prolong your suffering." And there was a certain thick-necked returning participant who'd taken the mattress closest to the door because he knew exactly how bad it was going to get. "I get it bad," he said. "Last time they made me leave the lodge and go out into the courtyard, then they kicked me out of the courtyard too." He did not say what bad meant, but he grinned in a way that made me not want to ask.

There were also the cases of psychedelic catastrophe that I'd read about before coming. The twenty-six-year-old man who "was seen to walk in front of a heavy goods vehicle while he was grinning." Or the two men in their twenties who each ingested several seeds of Hawaiian baby woodrose. One "experienced a sense of well-being as well as losing track of time." The other jumped out the window, falling several stories and suffering conquassation of the skull, multiple rib and pelvic fractures, lacerations of the right lung, cardiac contusion, haemopericardium, bilateral haemothorax, total rupture of the thoracic aorta, subarachnoideal haemorrhage, and atherosclerosis of the aorta and coronary arteries—death, in other words. A person was careful not to think about these things at this moment, but still, they were there in the mind. Somewhere out there, the spaceship hovered.

More worrisome than one's well-being during the trip, though, was the thought of one's well-being after. With any psychedelic, you're talking about the possibility of fundamentally altering the way your mind operates. Even for a person who's done their fair share of drugs, every trip is a sobering leap into the unknown. And ayahuasca in many ways is the psychedelic of all psychedelics. It's not the sort of thing where Jerry Garcia smiles up at you out of the carpet. It makes people violently ill. There's a lot of talk of snakes and pain and surrender. Any person getting handed the cup is bound to think very hard about what they've gotten themselves into.

"There's always the chance with psychedelics," the guy on the mattress next to me said. "You can get rid of something that turned out to be totally necessary to the function of your brain. This guy I know had a bad acid trip. He didn't realize there were socks balled up in the front of his shoes and broke all his toes."

"Wasn't anyone with him when he was tripping?" I said.

"No, no," the man said, turning his palms up in clarification. "This was twenty years later."

As the moment drew near when we would actually imbibe the liquid, these misgivings manifested in a mild but pervading mania. The mattresses were set out in two rows at intervals down the length of the lodge. Hardly more than eighteen inches separated one from the next, and in this confined space, each person suddenly set to arranging and rearranging the few belongings they'd brought along. They put their shoes at the foot of the mattress, moved their notebooks from their left side to the right and back again. They zipped their duffels, fluffed their pillows, folded their blankets, closed their eyes, and breathed deeply, then sat back up and checked to make sure their phones were off. (We'd been told the story of the panicked participant who got on her phone and called 911.)

"I don't know why I feel it's very important that my pen be exactly there," the man next to me said. "But I do."

The central focus of many of these rearrangements, of course, was the buckets. While we were being smoked with sage, a single bucket had appeared at the foot of each mattress. They were just white plastic trash cans, really, the sort you might buy in gross, each sheathed on the inside with a disposable white plastic bag. People tightened these bags over the rims, placed the buckets at their feet, their side, in their laps. They looked inside. Empty.

My personal quandary was my name tag, which was clipped uncomfortably to the neck of my sweatshirt. I was troubled by the image of the tag swinging down into my line of puke as I expelled my morning's Amtrak oatmeal, but in a more faux-philosophical way, I just couldn't wrap my mind around the idea of wearing a name tag on a psychedelic trip. Skeptical as I was of the idea of meeting earthen spirits, I didn't want to show up dressed like I was headed for the Javitz. I took it off and placed it on the floor beside me, but no sooner was it gone than I started to argue from the other perspective. What if I really did lose my mind? It occurred to me there was probably a reason each name tag had the name of the church and our specific place of lodging written on it. Which lodge was I in anyway? Hummingbird? Dragonfly? I saw myself, feet bare, pupils big as gumdrops, wandering the glass-strewn shoulder of some wide Floridian highway, grabbing the cashier at Dunkin' Donuts and demanding to know what was really in the eclairs. *If only he'd been wearing his name tag . . .*

I went to make a joke to the man beside me, but while I'd been dealing with my deep-seated name tag issues, he'd packed up his belongings. With neither explanation nor goodbye, he stood and left, and suddenly things were very real. Before I could think whether I should pack up my things too, the door to the lodge opened a final time, and there was some chatting at the far end of the room. A woman appeared silhouetted against the dusk, clutching a bag and water bottle.

"Katie?" someone said.

"That's not my name," the woman said.

"Your name tag says—"

"That's not my name."

And then Rayn appeared and took the mattress beside mine.

OUR FACILITATOR WAS NAMED SLEIGH, A SUBDUED but cheerful former therapist with thin hair and eyes that looked as tired and soft as shucked oysters. His steps were soft too, and careful. It seemed like a harsh word could fell him to the ground. In a throaty, frog-like voice, he told us we would soon be taking our medicine. In two hours, they would ring the gong, and at that time we would be allowed to have an additional portion, a booster. He strongly suggested, if we chose to take the booster, that we not simply take a quarter or a half a cup. He strongly suggested we consider at least a whole additional cup.

"I'm drinking half a cup tonight," he said, by way of encouragement, with a faint giggle. "And I'm leading this journey."

He instructed us to hold our cups to our hearts and think of our intentions. Then he pressed play on a phone, and the room was filled with a rhythmic quavering-alto-and-guitar-in-nature melody that can only be described as spiritual.

We could drink whenever we were ready.

In both texture and taste, it was indistinguishable from a pulverized mixture of molasses, Fig Newtons, and gravel. Like some bituminous excreta, it seemed to cling to the tongue and teeth, and even when we rinsed our cups with water and swilled this rinse water too the taste remained, somehow became clarified and stronger. Simultaneously cloying, grainy, and bitter, it did feel like medicine, a very strong medicine, the sort with which you might dose a sick pig, and this was disconcerting. One could not help but feel like the sick pig, who receives the medicine without ever grasping the nature of its ailment. In the half-light, you could hear people sucking at their teeth.

Sleigh instructed us to sit up for a little while and to wait to drink water. There was the possibility of a premature purge. Sleigh shook his head with sincere sadness at the thought. So, after drinking the brew, I scootched myself to the top of my mattress, nestled my spine into a spray-foam-coated cavity between a pair of vertical two-bys, and closed my eyes.

Considering that this moment marked my departure from the land of minute-hands and reason, it's probably a good time to pause and give a quick primer on how exactly Mother Ayahuasca works her magic. The main molecule you're dealing with in ayahuasca is this stuff:

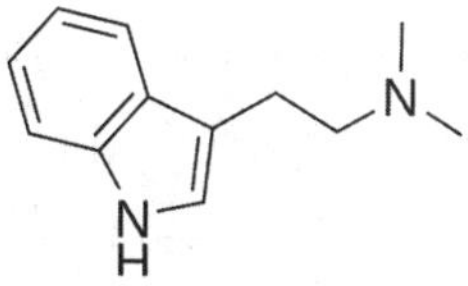

Harbin

It's called dimethyltryptamine, usually shortened to DMT, and it's part of a whole class of chemicals called tryptamines that includes psilocybin, lysergic acid, and others. You don't have to squint too hard at DMT to notice a distinct similarity in shape to this molecule:

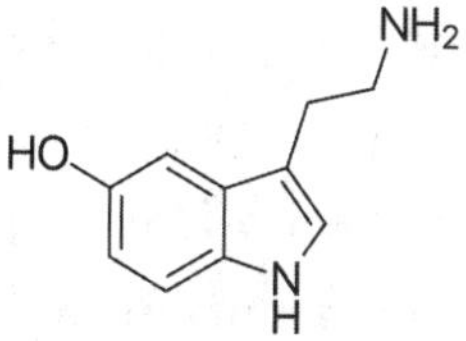

CLY

That's 5-hydroxytryptamine, 5-HT, one of the body's homebrewed workhorse neurotransmitters. You probably know it by its common name: serotonin.

Like so much about the brain, we have a general sense of what serotonin does (a lot), but we don't really understand how. It's a chemical messenger, basically a tiny molecular key, and it fits, as far as we know, at least fourteen different locks (the

5-HT receptors) throughout the brain and body. In humans it's involved in basic functions like moving the bowels, controlling the bladder, regulating body temperature, vomiting. But the chemical is better known for its work in the brain. Even though less than one in a million neurons make the stuff, the neurons that do are remarkable. In "a defined and organized manner" their tendrils infiltrate every section of the brain. Unlike regular neurons, which act basically like telephone wires, passing a signal from one end to the other, many of the serotonergic neurons in the brain are unmyelinated, meaning they have no sheaths. Instead, they act like soaker hoses, watering the brain with signal, in the process regulating everything from mood and perception to appetite, memory, and attention. "Indeed," as one article notes, "it is difficult to find a human behavior that is not regulated by serotonin."

Now, with DMT and all the other chemicals that in some way look like serotonin, what they do is selectively work, at much higher levels, on certain 5-HT receptors. In the case of DMT, the effects seem to relate mainly to 5-HT_{2A} and 5-HT_{2C}. What exactly they do there is anyone's guess. From EEGs, it seems possible that DMT alters the brain's hierarchy, "disrupting the normal state of top-down neural control and allowing greater bottom-up transfer of information." Meanwhile, MRIs show activity in areas of the brain involved in memory and visual processing. As for the subjective experience of DMT, it varies but has been described as following a distinct cycle. Half an hour after ingestion, you start to feel weird. Your perceptions begin to shift, you shake, you feel vulnerable. This is followed quickly by a period of acute paranoia, fear, and confusion, during which you sometimes have the terrifyingly real experience not of remembering but of reliving traumatic memories, which causes an overwhelming barrage of sensations and intrapsychic readjustments that, with a few bouts of intense vomiting, suddenly release you into "an expansive state." You encounter the spirits of animals and plants, make contact with higher powers, speed up, slow down, and travel through time at will, experience oneness with the cosmos, come to terms with death, crack open your pineal, know peace, etc.

As for my own experience, there isn't much to say except I had a wonderful time. No alien extended a hand from the door of its silver ship. I felt no impetus to vomit. For a bit, I just thought about my life and wrote in my notebook. At some point, finding it difficult to see the words in the dim light, I put down my pen and noticed that my hands were very large, comically so, and very still, like great wooden mitts. I marveled at them, rotating my arm at the shoulder and elbow so I could examine them without disturbing their stillness. Looking at them, I remembered how large and rough my father's hands had once seemed, and it occurred to me that those hands were now these hands. I also noticed—and this was the furthest reach of any hallucinatory phenomena—that by rubbing the tips of my thumb, index, and middle fingers together, I could engender a sort of meaningless phosphorescent dust. "That's interesting," I thought. Then I looked around and realized everyone else was already lying down, and so I lay down too.

I lay there thinking, mostly about my grandmother, whom

I'd loved. There was nothing particularly extraordinary about the experience except that it was some of the most pleasant time I'd spent in my own brain. I was aware, albeit in a distant way, that things were going on in the room around me. One of the volunteers in our lodge was the woman with her hair bleached white, and I listened as she padded back and forth across the room. I heard the door open and close with a whisk of metal on metal nearly identical to the sound of a samurai sword being drawn in a movie. "Shit," said the thick-necked man by the door who knew he was going to be kicked out. "Shit, shit, shit. Help." And I felt the vibration when his raised hand fell to the floor. From some corner came the sound of committed ralphing.

But then a sort of scuffling began to intrude on my line of thought, and I became aware that someone was doing something directly beside me. I opened my eyes and found Sleigh, or rather the white seat of Sleigh's pants, beside me. He was kneeling on the floor and leaning over Rayn. I let my head fall to one side and looked at her. Someone had given her an extra blanket, but even with this, she was shivering. Her eyes were pressed tightly together, and her breaths were short and forced. Sleigh knelt beside her.

"It hurts," she said, rushing the words out between breaths. "It hurts so much."

"I know," Sleigh said.

"Why am I shaking?"

"That's the medicine working."

VOMIT

We had been warned that everyone's purge was different. Some people might shake. And we had been instructed, as well, to keep our focus on ourselves, so I closed my eyes and tried to return to my thoughts. But through the thin wall, I could hear whispers outside, then the shuffle of many feet. When I opened my eyes again, Sleigh was gone and Sean, the medic, was in his place. Rayn's body had gone entirely rigid, and every few seconds a tremor passed through her limbs, banging her legs off the floor.

"Why am I seizing?" she said.

"You're not seizing," Sean said. "If you were seizing you wouldn't be able to talk to me." It was not clear if this comforted her. She had been holding her head up to speak, and now it collapsed onto her pillow.

"What's," she said, "happening?" Her lips trembled against her pillowcase.

A few other medical staff filed in then, and they laid a stretcher down beside her and rolled her onto her side while they slid it under her body. For a moment, she was turned so that we were looking directly at each other, but she couldn't see me. Then they took her away.

I closed my eyes again. I knew I should return to my thoughts, but it was difficult, not to be concerned that the person next to you was just taken away on a stretcher. I wanted to keep thinking about my grandmother, but I knew she would have been thinking about Rayn. The man in the corner had begun to moan more loudly too, which was making it hard to focus. "Shit," he said and spit in his bucket. "Shit." Someone helped him to his feet and out the door. "Thank you," he said. "Shit." And then they rang the gong.

Slowly, I sat up. Sleigh was already kneeling at the foot of my mattress, his fingertips tented delicately on the floor. "Booster?" He said it helpfully, glancing into my bucket as he spoke.

"I'm not sure I need one," I said.

He seemed pained by this response. "Are you receiving any messages?"

He looked at my notebook. I looked at my notebook. I thought about this. I didn't think I had received any messages, but I couldn't be sure. What qualified as a message, and how did I know if I'd received it?

"Is Rayn okay?" I said.

He looked away and pushed himself up from the floor. He seemed disappointed in me, and he mumbled something in reply—they were keeping an eye on her, I should focus on myself—but I can't quite remember it. I just remember that it wasn't yes.

I was about to lie back down when the volunteer came and knelt beside me.

"You're getting the booster, right?"

"I don't know if I need it."

"Listen." She tilted her head so her bleached white hair fell to one side of her face. "You came here to do some *work*."

"How long has it been?"

"Two and a half hours," she said, and then she lowered her voice. "I'm not even supposed to say this, but I've seen people who need a booster, and you need a booster. I've literally never said that to anyone before. What? You're writing in a notebook? Are you kidding? You should be on the floor. You shouldn't be able to move right now."

I laughed, but her face was serious. "Actually, you know, I noticed a lot of folks tonight weren't having super intense trips." She got up to go to someone else. "I think this batch might be a little weak."

And like that, as happens sometimes on any evening on Earth, the vital thread was mislaid. The vibes had shifted. I could feel it. I tried to assess the situation. I no longer felt calm. I could not stop turning over the words of the white-haired dancer. "She peer pressured me," I thought. "That was peer pressure." But I could not make up my mind whether this was a bad thing. I hadn't vomited, after all, and now I definitely felt like I was on the downhill side of wherever I'd been before.

I got up and went outside and found myself quickly sitting on a mat by the fire with another cup of ayahuasca in my hand.

I knew immediately that I didn't like it outside. There was, in drugspeak, a different energy. The moon was up, and the canes of bamboo were tilting ominously overhead. A few planes crawled across the blue-black dome, winking. People were scattered around the fire on mats, swaying side to side. A woman walked past with a blanket draped over her head like a shawl, the far corners dragging behind her on the ground. Her glasses sat askew on her nose, and squinting through them, she seemed to be using an empty mug as a divining rod. "Water," she said. "Water."

And an old woman appeared, her face so drained of life that I could not tell if she was the old woman I'd seen a few hours prior.

"I shit my pants," she said. "But two angels took my dirty drawers." Here she lifted her nightgown. "And they brought me these."

At the head of the fire, two facilitators, a man and a woman, both all in white, lay back on some fraying lawn furniture with a tray of cups set out before them and the brown goop in a Pyrex measuring cup, and every so often a person drifted toward them and received their sacramental booster. As I watched, Samvith emerged from our lodge, walking with a bobbing gait. The man stood and said something so quietly that Samvith had to lean forward to hear.

"Yes," said Samvith with a vigorous nod. "Yes, yes. Another please, a whole one." Then the man in white selected a cup from a tray on the table, stirred the sacred mixture with a tablespoon, dolloped some in, hummed with his eyes closed, whistled twice, and passed the vessel into the outstretched hands of Samvith, who scurried with it back to his mattress. I took a slug from my own cup and followed him.

It seemed to work more quickly now. Lying on my mattress, I had what passes, I guess, for a vision. I thought that as bees hurry from one flower to the next, humans dote upon one another's sorrow. We are drawn to the girl with the demented mother, the brother of the suicide, the woman with the scar on her wrist, the widower with three children, the child with dreams of wolves. We are drawn to them because in their suffering we recognize some image of our own. Pressure had been building behind my nose as I traced out this line of thinking, and at this thought, with a great gout of snot, a sob burst from me. There were no tears, just a quantity of mucus so voluminous that I had to cradle my face with one hand while I waved the other for a tissue. Sleigh dropped a few beside my mattress. My purge, apparently, had come from my nose. I blew it and blew it, and then I lay down and closed my eyes.

I WAS STILL LYING THERE WHEN I HEARD THE DOOR open and close. A shadow slipped across the room, and Rayn lay down on her mattress.

The music had ended by then. Most folks were asleep. But the man outside was still shouting and spitting in his bucket, and Samvith was sitting cross-legged on his mattress with his hands on his thighs and his back perfectly straight. His whole body was vibrating. He giggled nervously. He seemed to be arguing with something in the darkness.

Rayn was lying on her side, I could tell, looking at me. I rolled onto my side and looked at her. Her palms were together and her cheek lay against the back of her left hand. Maybe eighteen inches separated the tip of one nose from the tip of another. It was like we were at a slumber party.

And listen, it doesn't matter if either of us was in our right mind. What's a right mind, anyway? What matters is that I could see her face. She'd swept her hair back. There were no piercings, no tattoos. It was a long face, very elegant. The skin was smooth, almost waxy, and from either side of a narrow sloping nose, two deep-set blue-green eyes were looking out at me. It was the sort of face one sees at the bottom of a river.

"I'm glad you're back," I said.

"Me too," she said quietly. Her eyes narrowed on a spot in the air between us, then widened and settled on my face. "Why?"

"I was worried about you," I said. "That was scary."

"I was so scared." She closed her eyes and shook her head at the memory of it. "Did you get sick at all?"

"I thought I was going to, but it didn't happen. Lots of people did."

We paused and listened to the man still retching and cursing outside.

"I got so fucking sick. They took me to some room, I don't even know where." She shook her head again, as if to dispel the memory. "I don't know why anyone would do that to themselves. I'm just so glad to feel normal again," she said, and then her brow furrowed.

"What?"

"I just never felt that way before, happy to feel normal."

I looked at her and nodded and turned my gaze to the ceiling and its amorphous blobs of yellow foam. Somewhere, someone was hosing out buckets. When I turned back to her, she was still looking at me.

"What was it like for you?" she said.

"I don't know," I said. "I just sort of thought about my relationships, and then I got really congested."

I blew my nose. "What was it like for you?"

"I thought I was dying," she said. "I literally could not stop shaking."

"Did you think about anyone?"

"Not really."

"Shit," said the man outside. "Fuck, shit, fuck."

"But for a little while, it was like I was a kid again," she said. "It wasn't like I was remembering. I just felt like I did when I was a kid, you know, back before I wanted to die all the time."

The man outside seemed to have finally grown exhausted. He retched, breathed, spat. "Shit" echoed faintly in his bucket. But Samvith's argument was escalating. He stood now in the center of the room, slapping his chest and chafing his legs and speaking quickly into the gloom with the singsong cadence of Ahmedabad. "I am winning," said Samvith. "I am winning. I am having fun. I am having *fun*. Ha ha!"

"Why did you change your name to Rayn?" I said.

"I guess I was so used to people screaming my old name at me. I just didn't want to hear it anymore. And I liked Rayn. And you know, like, I spell it with a *y*. I thought it was really cool that way."

"It *is* really cool," I said.

Her nostrils flared, and the muscles around her mouth squirmed, and then without warning, she was smiling. "I know."

THERE WAS NO BREAKFAST FOR THOSE DOING THE morning ceremony, but it didn't matter. I wasn't hungry. It had been growing light when Rayn and I finished talking and drifted off, and I'd woken up shortly after dawn, but the lack of sleep had no effect on me. In fact, I felt filled with energy.

It was not terribly hot, but it was muggy that Saturday morning, and this, combined with the collective emotional interval training of the night before, gave the scene outside a feeling of slow motion. People walked unsteadily across the grass, and whenever they ran into another person, they stopped to talk. Little clots of people formed and separated and reformed, and conversations were springing up all over the place. Everyone wanted to know what others' experiences had been. Out by the medical tent, the old woman from the night before had been waylaid by a young hippie couple. Chin hairs shining wildly in the morning light, she was smiling broadly. "It was not being able to use my legs that surprised me. Thank heavens for that young man who helped me to the bathroom. I just wish his head hadn't kept changing. I don't know who all was giving him their heads but there was a lot of them."

By the empty firepit, a wiry middle-aged woman from Alabama with close-cropped hair was cheerfully putting the fear to Elsie, whose arms were wrapped more tightly than ever around her abdomen. "For me, it wasn't real like a hallucination," the Alabaman said. "It was more like colors, but the colors were feelings too. When I closed my eyes, all I saw was these mottled browns and mauves." In her pronunciation, the final word rhymed with *coves*. "I mean I don't mind browns and mauves usually, but these . . . I don't know how to explain it. They was just ugly to me, real ugly-like, I couldn't stand to see them."

"My husband," Elsie said. "He purged, but I didn't feel anything. When I got up to go to the bathroom, though, I was covered in shit." She was breathing faster as she talked. "I don't know whose shit it was, but it wasn't mine. It was *not* mine. I'm trying to figure whose shit got all over me."

"Just so ugly," the Alabaman continued. "Like worms or something . . . like a big old ball of worms just writhing around . . ." She paused while Elsie wiped at her shirtfront. "Like what you'd see if you was trying to dig out of your own grave."

The place was full of energy. Samvith stopped me and rubbed my shoulders. A volunteer carrying a towering stack of buckets told me a story about how he'd cheated on an old girlfriend and learned the chakras. When I got to the shower and ran my head under the water, I had the distinct sense bodily fluids were being exchanged in the neighboring stall. "There's nothing like puking together," as Plath says, "to make you into old friends."

I ran into Rayn on my way out to the morning ceremony. She was talking to someone official I didn't recognize, assuring them she had a place to stay in Orlando. She smiled again and held up two fingers in a peace sign.

"I'm out," she said.

"You going to be OK?"

She nodded.

I wanted to talk, but there wasn't time to hang around. The ceremony was starting.

"Take care of yourself," I said.

She looked very closely at me and said, "You too."

A few minutes later, I was preparing to take my second drink.

AS PROMISED, THE GROVE WHERE THE MORNING SES-sion took place was indeed magical. A few live oaks reached out across a central clearing, their limbs arching and overlapping and bending back down to the earth, creating a canopy of shimmering green shot through with sunlight. The women at the front desk had been effective in their sales pitch. Nearly the entire group had decided to undertake the extra ceremony, and we all lay scattered in this wavering dappled light, each on a worn rainbow rag rug. Even the ground felt remarkable. There was a light duff of leaves and pine needles, and the web of tiny oak roots gave the soil a light and springy feel.

It was about 10:00 a.m. and already almost uncomfortably warm, but there was a light breeze, and even if the shade wasn't dense, there was a delight in the way it lay in splotches across us. I was troubled by Rayn's departure. I felt some sort of bond had sprouted between us, and I was concerned about what lay ahead for her. But I had put it out of my mind for the time being. I felt good, and when a man in white asked what size cup I would like, I chose a slightly larger dose than I had the night before. I drank it down, and I diligently rinsed out my cup to get the last grainy morsels. I wanted the medicine. I was still nervous, but as I lay back on my rag rug with the drink inside me, trying to position myself as best I could out of the direct sunlight, I felt prepared.

They'd said the effects of the drug were cumulative, and as they cranked the volume on the weird pseudoethnic hippie music, I did become aware much more quickly not only of the derangement of my faculties but also of the degree of this derangement. It will give you some sense of my early state if I say that before long I lay on my side, deeply engrossed in the work of two ants. One was of the near-microscopic variety. It was traversing the back of my left hand, pausing every few steps to bow its head and bury its tiny and incredibly sharp pincers in my skin. The other I couldn't see, but I presumed from its weight and pace that it was one of the large red velvet ants I'd observed around my rug as I first lay down. It had crawled up the back of my arm and, after some hesitation, entered my shirt and begun to explore the terrain of my back. I did not feel I would accomplish much by trying to get rid of it, and anyway, I wasn't sure I could move. Besides, I was interested in this sensation of being walked upon, being explored, just as I was interested in the sensation of radiating pulses of pain and warmth as it sank some sort of venom into the small of my back. Many people were already puking by this time. The different varieties of ralphing echoed up from around the grove.

I began to notice my left eye. For years I had felt discomfort in that eye whenever I looked at a screen, and it occurred to me how absurd it was that I hadn't visited an optometrist. As I lay there, just feeling that side of my face, I began to notice a vibrating beneath the skin, or more like a writhing. There seemed to be a small cluster of mechanical maggots performing surgery in the muscles of my left cheek, I realized. I wasn't scared. I knew it wasn't real, although it felt very real. The operation was not so much unpleasant as novel. They were churning steadily through the cheek, moving in the direction of my eye, and I was curious what they would do when they got there.

There was much more vomiting then from the larger group. Cascading volleys of it. I looked around. The woman next to me was on her hands and knees with her face in her bucket. On the far side of her, Terrence sat rocking, holding his bucket to his chest like an infant. I supposed everyone reacted to the drug differently. Some people puked and others, like me, didn't. They just didn't need to.

Then I looked up at the leaves. They were swaying back and forth in the breeze, and they'd begun to swirl and pulse kaleidoscopically in time to the music. This sight brought with it no feeling of concern—I'd seen that kaleidoscope once or twice before in my life—but the swirling, combined with the deep bass of the music, did make me dizzy. I closed my eyes, but the dizziness only increased in time to the music, and the music had shifted, I realized. It was more rhythmic now, reverberating up through the soil, and it was dominated by some sort of shaker or maraca that sounded almost identical to the rapid shudder of a rattlesnake's tail. The rattling got louder, and the vertigo grew more intense, and then I thought that it was possible, maybe likely, perhaps even certain, that I was going to be sick.

You probably have experienced the creeping restlessness of nausea, and so you know its familiar but almost indescribable suite of symptoms. These were the sensations that began to overtake me as I lay there. My stomach didn't exactly hurt, but my cheeks were hot, my palms clammy. My throat was tense and my joints sore. More than all of this, though, there was a feeling of unwellness that seemed to go beyond the physical body. I flexed and unflexed my legs, stretched my jaw, breathed deeply, but I could not dispel this feeling of unease. I lolled on my rug, rolling my head around as I tried to locate my bucket. Finally, I latched on to it and sat up, but no vomit came, and soon, too ill to stay upright, I lay back down. The nausea just kept washing over me in waves, each one coming a little faster, cresting a little higher, than the last. I didn't know what was going to happen then—I felt no urge to vomit or to move my bowels—I only knew that the nausea was getting worse and I would soon be entirely incapacitated by it. Terrence's words returned to me: "I almost shit myself."

After the next wave had peaked, I gestured to the nearest volunteer.

She walked over to me, holding up her long white skirt as she stepped among the bodies. I don't remember her face, but I remember she smiled as she knelt beside me.

"I want to go to the bathroom," I said.

She nodded and helped me to my knees, and then to my feet. Then she looped her arm through mine and we began to walk, very slowly, picking our way between the buckets and the rugs. A few steps on, one of the long, drooping branches of an oak lay across our path.

"Watch out for that branch," I said, stepping over it with first one foot, then the other. And then my legs stopped playing the classical role of supporting my body. Everything went dark, but I could feel that a number of arms had caught me and were bracing me beneath my shoulders. I could hear several voices, and I could sense that I was being half-dragged and half-carried in some direction. There was a creaking as a door opened, a banging as it closed, and when I came to, I was standing in the rancid blue gloom of a porta potty. With whatever last reserve of energy and wherewithal I had, I undid my shorts, let them drop to the floor, and plopped backward onto the seat. A second later the door opened, a hand shoved my bucket through the crack, and it closed.

VOMITING IS EASY TO DISCERN, AND IT ISN'T HARD TO understand why it evolved (puking up poison confers a real advantage), but nausea's harder to pin down. Its evolutionary role isn't quite so clear. It's suspected it relates to memory—smell something that made you puke once, and nausea sets in before you can consume it again—but it's not even clear what exactly nausea *is*. The experience is almost entirely subjective, for one thing. The definition in the Oxford University Press's *Nausea: Mechanisms and Management*—"An unpleasant sensation of a protective mechanism elicited by the interaction of inherent factors and changeable psychological states"—is perhaps less a description than a testament to that description's impossibility. And it's maddening to study. Where does it reside, this feeling of unwellness? Or to put it another way: it's easy to tell how much chemotherapy a pig can take before it vomits. But what if it's a rat? What if it can't vomit? What's the limit of its suffering then?

These are important questions, or they became important to me then, because ayahuasca doesn't actually act on 5-HT$_3$, the receptor linked to vomiting. Instead, it takes the 5-HT$_{2C}$ receptor, which links to nausea, and jams its thumbs down on it hard. In most people, this causes vomiting, but as I sat there in the porta potty, I had a different experience. Some words came back to me, something a doctor had told me when I was a kid that I'd never really believed. I'd been ill, violently, life-threateningly so, and yet my body had refused to purge itself. "It's strange," the doctor had reflected, considering my crippling nausea but total absence of emesis. "You seem to lack the vomiting reflex."

It was very warm in the porta potty. Sweat poured off my body. It soaked my hair and my shirt, ran down my arms and my legs, dripped metronomically off my nose onto the hollow plastic floor. The temperature was only in the mideighties, but the humidity was high. The heat index, the actual experienced temperature, I later discovered was hovering around 102 degrees Fahrenheit outside. Take into account the greenhouse effect, along with the lack of air movement necessary to evaporate moisture, and the atmosphere inside that porta potty was probably closer to 110 or 120 degrees. I had unwittingly entered a sweat lodge.

I knew none of this then, of course. I just knew I felt very, very sick. With each wave, my body tensed, scattershot images from the night before appeared and vanished, until finally my brain ceased to function almost entirely and every frequency was taken up with the awareness not of pain—pain has a location—just nausea. Nausea in an ever-widening spiral, nausea and the firm belief that there was no way this particular sickness could do anything but go on forever. I felt I would have to strip myself from my body in order to escape it. I had been told to surrender. I wanted to. But to what? No light poured into my brain. No earth mother offered her dark and emetic embrace.

They say ayahuasca erases the line between mind and body, opens your third eye, strips your ego from your bones, your typical ration of psychonautical tripe, about as useful as saying nausea is the unpleasant sensation of a protective mechanism. But both the Oxford University Press and the dude talking chakras are taking different angles on a not entirely bogus point. The 5-HT receptors are tied up not only with feelings of nausea but also with your memories and your emotions, which is to say with your relationships and your feelings of self, and a feeling of nausea as intense as the one I then experienced is bound to do some psychic rejiggering at a very basic level. The idea, I think now, having spent a great deal of time trying to make sense of what happened in that porta potty, is for you to feel such an all-encompassing nausea, in both body and memory, that you have no choice but to puke, and in so doing to interpret nausea as something else entirely.

The trouble for me, I believe, was that my body couldn't quite figure out what that something else was. If I had been able to vomit, maybe I would have gotten there. Or if I had experienced the liquid magma–like diarrhea that the man on the rug next to mine later reported having. But there was no urge to do either. No saliva filled the mouth, no rumbling roiled the guts. I wanted to vomit, but no vomit came. There was merely this nausea, not in any distinct location in my body, but in *me*.

I don't know how long this went on. Sean, the medic, stood outside and checked in on me every so often. When I asked him later, he said only that it had been a long time. He'd been worried.

But then, little by little, the nausea began to ebb. The ability to control my limbs returned, and suddenly I was lucid, weak but lucid. It occurred to me that as long as I was in there, I might as well urinate, and I tried, staring down at my genitals silhouetted against the blue miasma below, but no urine came. When I stood, the floor was slick with sweat. My body was as slimy as an eel's. I picked up my bucket, pushed at the door with my shoulder, and stumbled back out into the grove. Though it was past noon by then, and brutally hot, the air outside felt cool, like falling into a stream. I retraced my path to my mat and sat down. The music had grown louder while I was gone, or so it seemed, and the sounds of puking had receded. I looked around.

Not far from me, a pale boy, one of the Joe Rogan acolytes, lay on his stomach with his face turned in my direction. His eyes were open terrifyingly wide, but they didn't see. He was covered in dirt. The earth around him had been cleared of twigs and leaves in the shape of a snow angel, and he was digging into the roots and soil with his fingernails, alternating between a stock queen lisp—"I'm so expressive"—and a growl—"No healing."

Terrence lay beside him, rocking and crying. "Joe Biden," he said. "Joe Biden never teach me shit."

On the other side of me the man in the football jersey lay on his stomach. Every two seconds, he dolphin-kicked the ground, stroked his chin, and let out a Sherlock Holmes–style *hmmm*. Dolphin kick, chin stroke, *hmmm*, dolphin kick, chin stroke, *hmmm*. It did not end. Maybe it was this that made the woman directly next to me laugh and let loose a gush of urine.

And slowly my eyes rose and took in the facilitators and volunteers who were supervising the event. They were all in white, as usual, and some were weaving among the prostrate bodies, waving burning herbs and shaking rattles and singing along, eyes closed, to the music. Others sat at the periphery in chairs, faces sealed in beatific smiles. And it occurred to me as I surveyed this scene that we'd been eating low-calorie diets for a week, had subsisted on little more than celery and sweet potato chips for the last twenty-four hours, had been separated from our loved ones, deprived of intimacy, of sleep. All this, and then we had been given one of the world's most powerful psychoactive substances and set out for several hours on a hot summer day in central Florida. And I had my only real epiphany of the trip, albeit one I should have had much earlier.

"Oh my god," I thought. "This is a cult."

And then a lesser but more urgent realization: flee.

I scribbled a few lines in my notebook, glanced over my

shoulder to the road. I could see the whole sequence of steps. Call the taxi, gather belongings, wait by the front door, but even as I jotted it down, it began to seem less feasible. The joints in my hands felt sore, I noticed as I wrote. Whatever adrenaline had carried me out of the porta potty was waning now, and as I looked around, I began to feel dizzy again. I went down unwillingly, first onto one elbow, then onto my side, then flat on my back, and for a long time, I lay like that, a sick dummy in the woods.

IT'S HARD TO SAY EXACTLY WHAT MY STATE WAS when they rolled me onto the stretcher and loaded me on a golf cart. The man in the jersey was still dolphin-kicking, chin-stroking, hmmming, but most of the others had begun to roll up their mats and head back to their lodges. I was conscious, but barely so.

The memories that follow are as clear as if they just occurred, but they're separated by empty spaces where I'd be hard-pressed to explain what exactly happened. I remember trying to make some gesture with my hand, but I could not seem to lift it. It hung weird and dead from the wrist. And I remember that as I was carried into the medical room, I was very cold. I'd cleared the drug—they checked to see whether my pupils retained the telltale dilation—but I was so dehydrated it was hard to tell the difference. As I came to, the medics were sorting through their kit. An amorphous darkness in the corner of my vision slowly took on substance, and I turned my head to find myself staring at a jaguar tapestry. I had enough sense not to ask that it be taken down, but I did not look at it directly.

"I'm just looking to see where we can get this IV in," Sean said, wrapping a tourniquet around my bicep and rubbing the inside of the elbow with alcohol.

"I've got good veins," I said.

"Sure do, buddy. Problem is, when you get as dehydrated as you are, it's a little harder to get a needle in there. I'm going to count to three. One, two—" He jammed the needle home.

"You did the thing," I said.

"Where I put it in a little before three?" he said. "Yeah, I'm sorry. I lied."

"No," I said, closing my eyes. "It's okay. It's a good lie."

WHEN I OPENED MY EYES AGAIN, SEAN WAS SITTING at a chair by my feet. Another medic sat beside the bed. He had the meat-and-potatoes physique of one of my uncles, a farmer, and he seemed to be looking out across a field of wheat. I asked him his name. Tom, he said. He was from Sarasota. The room was quiet. A man came in to rummage through a duffel bag in a corner. A man came in to request a boutique IV, and when they went to check the veins in his arm, they found the crook of the elbow covered in an elaborate tattoo. "I got that to remember my boy who died," he said. "He took his own life." A woman came in looking for some paperwork. A woman came in with a question about her toe.

"I think I might have broken it during the ceremony? Is there a way to tell?"

"Oooh," said Sean, looking at her feet. "I see it right there, yep. The only way to be sure if it's broken is to go to the hospital and get an X-ray."

Tom looked up from the corner. "But they're just going to tell you not to walk on it."

"They can't do anything to fix it?"

"Nope," he said. "Just got to give it time to heal."

"Huh." She looked down at the toe. We all looked at the toe, which was bare on the cold linoleum, its nail rimmed with dirt, a toe that might or might not have been broken but anyway couldn't be fixed. "But you think it's broken?" she said.

I closed my eyes again.

"Well," said Sean. "Which way does it normally point?"

When I opened my eyes, Sean was looking at his phone. "Shew," he said. "I forgot."

"Doesn't feel like twenty years, does it?" said Tom.

"I still remember where I was."

"Everybody remembers where they were."

"What?" I said.

"I was just saying I forgot," Sean said.

"Forgot what?"

He put his phone away. "I keep forgetting," he said. "It's 9/11."

I had known this. You don't plan a puke-weekend in Orlando on the anniversary of 9/11 without it crossing your mind. But I'd forgotten too. That's the way 9/11 is, despite the whole *never forget* thing. You do forget. Then you remember again. And that's sort of the point. That sudden wallop of remembrance.

"I was just getting off duty," Tom said. His chest rose and fell. "The news came over the radio in my car."

"After the second plane hit," said Sean, "I told my wife I wanted to join the Marines. She said, 'I'll divorce you and take the kids.' Then I said I wanted to be an EMT. 'I'll divorce you and take the kids.' Of course, if I'd known where things were going . . ." He reached out into the air above my feet and signed a sheaf of invisible papers with panache.

"Bush was in Sarasota that day," said Tom.

"I remember," Sean said. "I remember watching his plane take off, and it just went straight up." He ran his hand vertically into the sky. "I remember just praying it got up there. I was so scared a missile was going to hit it."

We were all silent for a while. I don't know what they were thinking about. They seemed to have forgotten I was there.

"He was reading to kids, wasn't he?" I said. "That's what Bush was doing when they told him."

Tom turned and for the first time looked at me. "That's right," he said. "He was reading to kids."

"They came and they told him," I said. "But he kept on reading."

"Yeah," said Tom. He sounded puzzled. "He kept on reading."

"I used to hate him for that," I said. "But it doesn't seem so bad anymore."

Sean went out to get some food then, and it was just Tom and me.

"I'm sorry, Tom," I said. "I'm getting out of here. I'm spending tonight in a hotel room."

I thought briefly that my sense of the passage of time had become distorted again because for a long while Tom just sat there, motionless, considering what I'd said. Then he nodded.

"What brought you here?"

It's a hard question to answer, even if your mind hasn't been unwillingly stripped down, disassembled, and hastily puttied back in place. I thought about Erin asking me what I was trying to accomplish. I thought about Phil's hands when he phones home to his wife and Samvith's semi as he queues up another episode of Joe Rogan and the thumb Rayn puts out at the side of the road. I looked up at the ceiling, but I seemed to see through it, up to the sky, where Air Force One was still shooting straight up like a rocket.

"Tom," I said, "I have no idea."

He shrugged.

"You were called."

EVERYONE HAS SOMETHING THEY CAN'T LIVE WITH- out. Whether it's benzos or booze or the sound of a name screamed at nightfall and it's your name and it always will be. They say ayahuasca can help. It can clarify your feelings, it can relieve you of sorrow, though that relief might be made possible by the destruction of what you think of as *you*. "I've seen healings," Sleigh whispered to me that first night in his frog-like voice. "I really have." By that he meant he'd seen different people leave than the ones who came in. I believe him. I get it. Ayahuasca isn't truth, but who wouldn't want to live in the world it promises? I want to live in a world where Rayn starts her own tattoo shop and Phil runs his hands through his wife's hair. In my notebook, two words are scrawled across a page, written just before my fingers stopped working. *Never again.* A strange thing to write, when you think about it, the sort of thing you'd only write if you thought one day you might forget the mistakes you once made.

XI

Hair

> he did putt his hand vppon the french bodyes, and that heerevppon the said Margaret asked the said Thomas (meaning the articulate Thomas Follett) whether he were abasshed, and that vpon this speeche he catched her by the wrans nest, and plucked awaye somme of the hayre and wrapped it vp in a paper & shewed it in one George Lanes house in Loopitt

From a deposition in *Follett v. Stone and Tottle*, a 1614 defamation case in the Consistory Court of the Diocese of Exeter revolving around the theft and subsequent display of several pubic hairs

---

Terms

n. mop, rag, moss, down, do, locks, tresses, coiffure, curls, ducktail, beehive, Afro, flattop, fur, pelt, pubes, mane, scruff, cut

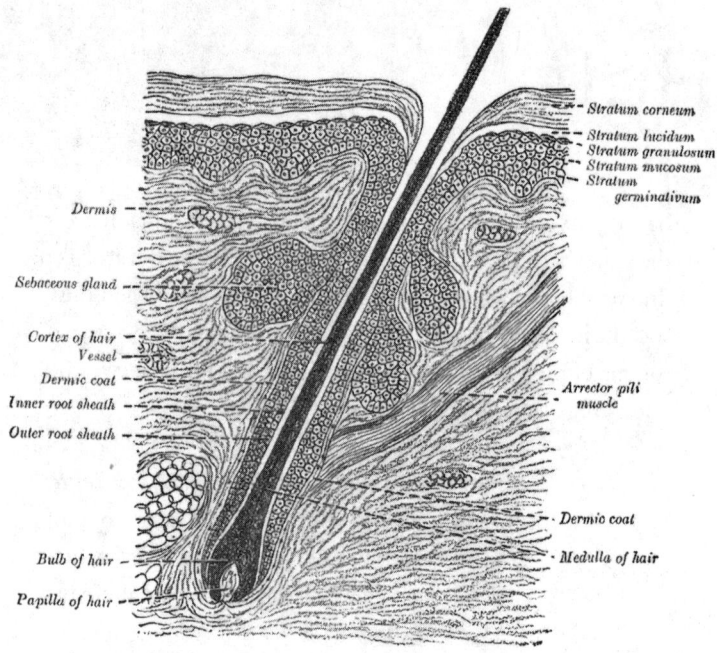

Illustration of hair follicle, *Gray's Anatomy*,
Henry Gray and Henry Vandyke Carter, 1858

Biological Prologue: Hair Structure

With the exception of a few small areas—the soles of the feet, the back of the ear, the palms, and the lips—the entire human body is covered in hair, about two million all told. Much of it, at least initially, is so thin and soft that we hardly notice its existence. It's called vellus hair, and it helps to disperse oils across the skin and regulate the body's temperature. The hair that most people think of when they think of hair is called terminal hair, and in children, it's mainly found on the head. It's confined to the scalp until puberty, when the circulation of hormones in the body, mainly testosterone and dihydrotestosterone, begins to transform many vellus hairs into terminal hairs. About 30 percent of the surface area of a woman is covered in terminal hair, mainly found on the head, groin, and armpits. In men, terminal hair can be found almost everywhere, covering as much as 90 percent of the body.

Hair follicles contain some of the fastest growing cells in the body (which is why chemotherapy, which affects rapidly dividing cells, tends to cause hair loss), and human head hair grows roughly twelve centimeters, or four and a half inches, each year. That equates to about ten micrometers every hour per hair, so that, when you take into account the average hundred thousand fibers on the human head, you grow about a meter of hair on your head every hour. This rate is fastest in early spring, while the rate of hair shedding is highest in fall (about sixty per day). But not all of the hair on the body grows at this same rate or on this same schedule. A hair on your head will grow for anywhere from two to seven years, while your eyelashes go through a roughly four-month growth cycle. The beard, for unclear reasons, seems to grow fastest in July.

Hair is, more than anything, a function and a signifier of the body's hormonal state, and because hormones are a function of our age, hair is intimately connected with maturity. The diameter of our hair increases as we leave childhood, peaking in the midtwenties in men, possibly later in women, before beginning to decline, giving the hair of recent college graduates a distinctly thicker, bushier appearance. And as we continue to age and hormonal balances shift, hair begins to take on different aspects. The drop in estrogens begins converting the fine vellus hairs on the upper lips and chins of women into thicker, longer terminal hairs, and in men, the hair follicles on the head begin to atrophy under the effects of too much dihydrotestosterone. Many men—science cannot say precisely why—begin growing hair inside their ears and nostrils.

The color of hair derives from flecks of melanin embedded in the strand. They're produced by melanocyte stem cells within the follicle, eumelanin for dark hair and pheomelanin for lighter hair. Most people have black or very dark brown hair, with brunette (~10%), blond (~2%), and red (~1%) making up the remainder. The original hair, it's theorized, was dark and curly, and lighter pigmentation and straighter locks were the result of genetic drift in the absence of negative selection following the migration of humans out of Africa and into more northern latitudes. Though blond hair is often associated with Northern Europe, genetic analysis has revealed the gene first developed about seventeen thousand years ago around Lake Baikal near the border of Russia and Mongolia and was brought to Europe in a long-forgotten mass migration from the steppes.

Red hair, meanwhile, is stranger and more ancient, possibly because the mutations involved are simpler. It stems from a change in the MC1R gene that affects the production of certain proteins in melanocytes. Because MC1R affects the pigmentation of the skin as well as the hair, those with a mutation of the gene are also more sensitive to sunlight. In fact, they are so much more susceptible to burning from ultraviolet light that having red hair is equivalent to an additional twenty-one years of sun exposure. Strangely, since one of the MC1R proteins is involved in both the perception and the blocking of pain signals, those with red hair also perceive pain in a fundamentally different way, are more sensitive to opioids, and tend to have higher pain thresholds. While red is currently the rarest hair color, a sizable proportion of those in the United States carry the gene recessively, and similar mutations have long existed. We now know, for instance, that some Neanderthals had red hair.

One of the most notable effects of aging upon hair, of course, is the loss of this pigmentation, which follows a roughly 50-50-50 rule (50 percent of the population is 50 percent gray by the age of 50). What happens is the melanocyte stem cell pool that endows hair with its color dies, and the result is a different kind of hair altogether: unpigmented, thicker, more susceptible to weathering and UV damage, and therefore frizzier and more fragile. It often grows more quickly, as well, as evidenced by the unruly eyebrows of many older men. Gray beard hairs grow up to four times faster than pigmented ones.

While it might seem shiny and lustrous in commercials for shampoo and conditioner, hair is not smooth. Seen with a scanning electron microscope, a single fiber of human hair looks more like the trunk of a hickory tree. The cuticle, the outermost layer, is composed of flat, overlapping cells like scales, and these jagged structures would immediately reduce hair to a tangled mess if a layer of fats, mainly a lipid called 18-MEA, were not bonded to their surface. It's this fatty layer that endows hair with the qualities we often think of as beautiful—the luster, the shine, the softness—and this layer of lipids is also the reason you really shouldn't wash your hair: 18-MEA is broken down through environmental wear and tear, and it is especially degraded by alkaline solutions like those used in many shampoos, particularly in children's no-tear formulations. Once this layer has been removed, friction increases between the strands and the hairs themselves more readily absorb and lose water, becoming brittle in the process.

Running through the middle of each strand, the melanin-flecked cortex gives hair its strength, body, and color, and when cut into a cross section, it reveals a structure similar in principle to the braided steel cables that hold suspension bridges aloft. Just as a single steel cable is made up of several braided ropes, the cortex is composed of spiraling macrofibrils, and as each of those braided steel ropes is itself composed of strands of twisted wire, nested within each macrofibril are spiraling microfibrils. But hair carries this logic further, with spiraling tetramers nested within each strand of microfibril and spiraling keratin proteins within each strand of tetramer. (Curl, by the way, is a function of this internal structure and of the angle at which the hair sprouts from the skin. Straight hairs have a symmetrical round cross section and emerge at a ninety-degree angle, while curly hairs are asymmetrical, ovoid, and emerge at an angle.) The comparison to steel isn't flippant, as human hair has an almost identical tensile strength to traditional steel. You could easily suspend the Golden Gate Bridge with hair, provided you could grow it long enough.

Interestingly, despite having lost the thick, rough coats that many mammals bear, we still have a reflex common to our pelted cousins. In cold weather or moments of fright, small muscles at the base of each hair, the arrector pili, tighten, causing the hair to stand up straight, helping to insulate the body or making the animal appear larger to intimidate rivals and predators. Though largely vestigial in humans, we do retain the reflex, so that the chill breeze or the raised knife in a horror film causes the hairs on our arms or the back of our neck to instinctively rise. Weirder still is that these arrector pili muscles, due to their minuscule size, are some of the first muscles to experience rigor mortis, so that before the body stiffens, its hairs first stand on end, giving to the recently deceased the appearance of having been surprised by death.

Nest

I have a cousin with good hair. It's fine and golden, and he styles it so it shoots up from his forehead in a bouffant wave. It's the kind of hair that would look great on a yacht off Nantucket, and he knows it. We get along, perhaps better than most folks in my family, but we're very different people. He drives a new Audi, wears freshly pressed Brooks Brothers button-downs, will probably one day own a wristwatch by Patek Philippe. I drive a dented old Toyota and am happiest in flip-flops and a short-sleeved tee. When I wake up in the morning, I take a cup of coffee out to the garden. He checks his brokerage accounts. We don't hide our points of view from each other. He thinks, rightly, that my hippie act merely obscures a deep essential vanity. I just always thought he was shallow. Or I did until he said something a few years ago, after the death of his mother, my aunt. Sorting through the endless clothes my aunt had left behind, my mother brought one of the dresses to her nose.

"It doesn't smell like her anymore," my mother said.

"No," said my cousin, "but her hairbrush does."

One of the notable attributes of hair, as Montaigne noted in his *Essais*, is its miraculous power of retaining and communicating smells:

> He who complains of nature that she has left man without an instrument to convey smells to his nose is wrong, for they convey themselves. But in my particular case my mustache, which is thick, performs that service. If I bring my gloves or my handkerchief near it, the smell will stay there the whole day. It betrays the

place I come from. The close kisses of youth, savory, greedy, and sticky, once used to adhere to it and stay there for several hours after.

But it's not just that hair holds on to smells, though it does. Somehow it also alters or mixes with each smell, changing it, deepening it so that the smoky residue of a campfire or the sweet earthy fragrance of coconut oil is wholly unique to the head of hair in which it is trapped.

It isn't clear to me why the smell of hair is so vital and elemental, far more so than many of the body's other smells. Perhaps because in their hair the person carries with them some olfactory record of where they have been and what they have done, or maybe, as with Montaigne, it draws some power from the association with love. How often, after all, do we bury our faces in the tumult of another's hair or sweep a stray strand out from between the lips of a loved one? How many times did I fall asleep in front of the television on my grandmother's floor with my face buried in one of the pillows from her bed? How often have I been transfixed by the smell of sunscreen, tempered by sweat and the heat of the sun and fixed by the brine of seawater, in my daughter's hair? At moments like these, like the pharaoh in the ancient "Tale of Two Brothers," whose clothes are washed in the tide where a lock of a woman's hair is floating, I am spellbound by the scent.

Even though they would have despised each other, I think my cousin has something in common with Jonathan Swift, who despised almost everything, hiding himself away each year on his birthday and meditating over the third chapter of Job:

> Why did I not die at birth,
> Come forth from the womb and expire?
> Why were there knees to receive me,
> Or breasts for me to suck?
> Now I would be lying down and quiet;
> I would be asleep; then I would be at rest
> With kings and counselours of the earth
> Who rebuild ruins for themselves,
> Or with princes who have gold,
> Who fill their houses with silver.

It was Swift who had deeply engraved and gilded upon his tomb a message letting passersby know that he had gone *ubi sæva indignatio ulterius cor lacerare nequit*, where savage indignation could no longer lacerate his heart. I can't imagine my cousin leaving behind such a dour epitaph, but I still think there is some commonality between the two, especially when I consider what was discovered in Swift's desk after his death: a lock of hair, presumed to be that of Esther Johnson, the woman he loved most in the world, with a note written in his hand, "Only a woman's hair."

I find this lock of hair, without name or date, this pathetic anonymous laconic pseudo-eulogy, more affecting than all the elaborate hairwork memento mori of the Victorians combined.

The gilt mourning brooch with its fleurs-de-lis of dun tresses, the faded daguerreotype wreathed in black curls, these seem corny by comparison. Here a woman once stood. She laughed, she scowled, she made of the evening a delight, she had a name, a way of tilting her head when she sang, and now all that is left is a lock of hair. The note's cynicism lends it its beauty. To recognize the utter lack of value in a thing and to treasure it still above all else, the heart breaks to bridge these two truths. People thought Swift cold, but a cold man would have thrown the hair away.

Why do people keep hair? Maybe it's just because the braided cable-like structure of keratin resists microbial degradation. Hair lasts, and so we keep it, whether in envelopes or lockets or specially made flowered china vessels. Or maybe it just seems a little cleaner than the other things we might preserve. Or maybe it's because we make a lot of it—thirty or forty feet—over the course of a life. Whatever the case, we do gather it. We buy Elvis's hair and Justin Bieber's. Through the temperature-controlled auspices of the Smithsonian, we keep a collection of locks supposedly belonging to the Founding Fathers. After Poe died in Baltimore, following a "vacant converse with spectral and imaginary objects," his body was "visited by some of the first individuals of the City, many of them anxious to have a lock of his hair." And when my childhood dog, a golden retriever, had to be put down, after I'd buried my face in her fur, weeping, as my aunt drove us to the veterinarian, the vet surprised me by returning with an unrequested ziplock bag containing a tuft of that tawny fur, which years later, discovered in the midst of packing my belongings for college, still retained its power and gloom.

Maybe an essential part of our personality does live in our hair. A person's hair can tell you a great deal about them, after all. The shorn locks of a Hindu man tell you that he is in mourning, and like a wedding ring, the beard of the Hasid tells you that he is married. From the tonsure, we know the monk has devoted himself to God, much as the dreadlocks of a Niyabinghi man declare his belief that Jesus returned to Earth in the body of Emperor Haile Selassie I of Ethiopia. The mom cut, "a soft waterfall in the front, but knives in the back," acts as a letter of introduction to the society of suburban motherhood, while the whole of the Nancy Kerrigan saga is captured in the crisp curl of bangs on young Tonya Harding's forehead. And you will understand a crucial datum about me when I tell you that, due to the similarities in our hairstyles, for a portion of my childhood in central Pennsylvania, I was nicknamed Tina Turner. "No one wants the cut," as Vanessa Bayer says. "The cut chooses you."

To be stripped of your hair is to lose some part of who you are. Maybe this is why my father has had his hair cut by the same person for nearly thirty years. He doesn't trust anyone else. Maybe it's why a balding friend of mine finds it embarrassing that people can see his scalp. It's also why there is always something uncanny about a really good wig. Even when they are removed, they still seem just slightly alive. For instance, I remember that after my aunt was diagnosed with cancer, she received a series of boxes from one of her friends, a woman who'd recently finished her own round of unsuccessful chemotherapy, and inside each box was a different wig. Well-made and stylish wigs not being cheap, my aunt then wore these over the following months, and continued wearing them even after the death of the friend who'd gifted them, so that the friend's son, seeing my aunt one day on the street, was momentarily paralyzed.

Of course, many men are mortified at the thought of losing their hair, feeling that it symbolizes their youth or their virility, though it's hard to see too much wrong with a biological occurrence that encompasses the lives and aesthetics of both Siddhartha Gautama and Jean Genet. Still, a lot of men feel they must hide the loss of their hair. Sinatra had a toupee collection. Burt Reynolds owed close to $122,000 to his toupee maker when he declared bankruptcy in 1996. Andre Agassi felt so defined by his youthful mane of blond hair that he lost the 1990 French Open Final, having spent the night prior frantically trying to repair his disintegrating wig.

These days, if he fears the loss of his hair, a man can simply pop a 1 mg pill of finasteride, the 5α-reductase inhibitor better known by its brand name, Propecia. A friend of mine with suspiciously full hair told me about the discovery of finasteride, how it was originally in trials as a cancer drug, to which end it totally failed—"They've got patients dying left and right, man, but then the researchers start to realize *all these dudes are dying with these full heads of the most luscious beautiful hair.*" It was a striking image, ranks of hospital beds, each one containing a withered man with gorgeous hair, as if the hair had somehow sucked the life out of the person it adorned, and I wondered whether my yacht-bound cousin, given the choice, wouldn't rather die with good hair than live with bad.

I didn't learn until years later that, while perhaps there had once been some original seed of truth in my friend's story of the cancer trials, the tale was in the strictest sense entirely inaccurate. Finasteride had never been trialed as a cancer drug. Its development, in fact, was far stranger. At a 1973 meeting in Atlantic City, New Jersey, an endocrinologist from Cornell described a remote village in the mountains of the Dominican Republic. An unlikely gene mutation had surfaced in this population, causing a portion of the children to be born intersex. At birth, the children appeared female and were raised as such, but at puberty, their bodies became muscular, their voices deepened, and a functional microphallus emerged. In the village, they were called the *guevedoces*, literally "penis-at-twelves." Their unusual development was found to be the result of a deficiency of the enzyme 5α-reductase, but to the scientists at Merck who eventually learned of the population, there was a far more miraculous aspect to the *guevedoces*: they knew no male pattern baldness.

There is no magic in hair. But it hardly matters, so long as we believe otherwise. There will always be something ineffable about a human's hair. Maybe this is why, whenever scientists seek to communicate the eerie precision of their instrumentation, they always cast it in terms of the width of a human hair, so that we know the Hubble telescope can see a human hair two hundred meters away, and the structure of the James Webb

Space Telescope retains its shape to a tolerance of 1/10,000th of a human hair, and the LIGO observatory can measure the distance to the nearest star system, Alpha Centauri, to within the width of a single human hair. Maybe this is why Edison, before settling upon carbonized cotton thread for the filament in his famous lightbulb, first trialed a strand of beard hair. Perhaps this is why the rumor persists, entirely unfounded, that for a time after we die, our hair keeps growing.

Somehow hair never stops being alive, even when we're dead. That's why the hair of Emperor Tewodros II, long held in London's National Army Museum, was repatriated to Ethiopia for burial nearly 150 years after it was taken as a souvenir following the Battle of Mandala, and it's why the still-luxuriant tresses of Queen Tye, dead three thousand years, inspired such awe and attention upon their display in Egypt in the spring of 2021. And it's this dead-aliveness that sent the Content Committee of the United States Holocaust Memorial Museum in Washington, DC, into a years-long dispute over how to exhibit nine kilograms of human hair of the roughly seven thousand kilos discovered by the Red Army at the liberation of Auschwitz, hair that had been intended for the manufacture of "hair-yarn socks for U-boat crews" and "hair-felt stockings for employees of the Reich railways" as well as for textiles in the interiors of German automobiles, hair that much of the committee hoped to display as dramatically as possible, in the form of a wall of human hair, in order to drive home the notion that the human body itself, in the eyes of the Nazi, was nothing more than matériel.

The original seven thousand kilograms had already been on

display for some time at the Auschwitz Museum in Poland, in large heaps of braids and knots and tresses and waves inside a row of glass display cases on the second floor of Block IV, and the Holocaust Museum had determined that, lacking human cells, the exhibit would receive no rabbinical objection. But when the committee met to discuss its display, "it became clear," as one staff member said, "that the members viewed human hair differently."

"While we recognize and share with you the concern for a means to convey both dramatically and soberly the enormity of the human tragedy in the death camps," read a staff memo composed by a number of dissenters in early 1989, "we cannot endorse the use of a wall of human hair." It would violate the innate sanctity of the hair, they felt, and it would cause in visitors a response not so much empathetic as ghoulish. Which was true, but it was precisely this ghoulishness of the exhibit that drove the opposing perspective. As one Auschwitz survivor said, "There is nothing that speaks louder against the Nazi crimes than this hair." Which was also true. The hair, for one of the directors of the planned exhibition, would be the "crescendo." "What is hair for most of us?" said another director. "It's our mothers, it's our lovers, it's the thing we come close to, a spot we nestle into." You could see photographs, shoes, suitcases, identity cards, cans of Zyklon B, could even enter a model of the gas chambers, but it was not until you were confronted with the hair of the murdered that you began to comprehend the awfulness of what had occurred. How could the Holocaust be felt, he argued, without this "sea of hair"?

A year after the initial memo of dissent, the Content Committee began to discuss the possibility of a privacy screen for the hair, to protect the display from the "casual visitor." The hair would be in some sense clothed this way, and a museumgoer would be confronted only if they chose to be. But by October of that year, the committee had shifted further. Any display of hair was in peril, though the director still held out hope. He dispatched a photographer to Auschwitz to photograph their display, with the aim of convincing the committee of the hair's power.

It was all a relatively moot point anyway, since the head conservator at Auschwitz was already struggling to prevent the seven thousand kilograms in his charge from completely disintegrating. The hair had long been stored without temperature or humidity control. From time to time, it had been washed, and at some point, curators noticed that moths had made a home in the hair. Since then, it had occasionally been spread on screens and suspended over trays of liquid naphthalene, the coal tar distillate used to make mothballs, so that the gas from the chemical imbued the hair with a scent repulsive to insects, but while this process had succeeded in preventing infestation, it had also left the hair so brittle that even the staff were afraid to touch it. As the conservator at the Auschwitz Museum said at the time, surveying the fragile tresses, "No one has ever had this problem before."

The crux of the argument against display was never clear. Some were worried moving the hair from its original context had changed it from an artifact to an exhibit; outside of

Auschwitz, it could only be viewed voyeuristically. Others were concerned it did not befit the National Mall to display the hair of the murdered. A theory emerged that it would cause in visitors to the nation's capital the exact sort of anesthetization of the empathetic faculty that had fostered Nazism in the first place. And running deep beneath all these threads seemed to be an inchoate feeling that simply to show evil was to become its apprentice. It wasn't even clear from day to day exactly who did and didn't want to display the hair. In the middle of the dispute, one woman in favor of display abruptly changed her mind. She had realized the proposed wall of hair might contain her own mother's locks.

And this was what my cousin and Burt Reynolds and the dissenters on the Content Committee got right: the hair *is* the person. Before my aunt died, she lost her hair. I hadn't seen her in months. I'd been living in South America, and when I returned home I found her sitting on her back porch. I had known she was sick, had known about the surgeries, the radiation, the chemotherapy, but her illness up to that point had been far away, unphysical. As I came around the corner of the porch, I was unprepared for the gaunt figure I discovered there, sitting in a wheelchair with a blanket across her lap. A thin hand feebly turned the pages of a magazine, but she did not read. Her hair was gone, or not entirely gone. Her head was encircled by an orangish fuzz, like the prickles you see on a baby orangutan's scalp. Only a single lock of her old hair remained. From a place above her right temple, it hung down, pale and brown, tracing the line of her jaw. She was so unlike the woman I had known and so like a skeleton that I froze. She looked up then, and saw me, and smiled. I had surprised her, and she reached up instinctively, as if to tidy her appearance, and tucked that final strand of hair behind her ear. It was a vain gesture, almost girlish, but it was the last gesture I can remember that was truly hers.

To die, as a shrine in Tutankhamun's tomb describes, is to comb the hair of Osiris. It's a beguiling image. Even the unbeliever recognizes the profound comfort and intimacy in the act of brushing another's hair or having one's hair brushed, the way it seems to return us in an instant to childhood.

I have been fascinated by the ancient Egyptians since around the age of twelve, when my nickname was no longer Tina Turner but Carlos, after the quick-footed blond-Afroed Colombian midfielder Carlos Valderrama. Back then I built a scale model of the Pyramid of Khufu as it would have looked in the Fourth Dynasty, with its polished marble sides and its point wrapped in hammered gold, and a special panel that folded back to reveal, suspended in midair by fishing line, the interior passages and chambers. I'm smitten by their hairstyles and their makeup and their keen obsession with death. At dinner, wearing their hair long down their backs or donning wigs in the "double style" with a curled upper section surmounting a cascade of long narrow plaits, ancient Egyptians would sometimes have a little model of a sarcophagus brought around to each of the guests in the hope that, by reminding each person of their looming demise, the company would be incited to greater revelry. I'm smitten too by their inexplicable passion for the matryoshka-like nesting of things one within another, so that Henutmehyt was found within three gilded coffins, and Merneptah within four, and Tutankhamun was entombed within a stone sarcophagus, two gilt wooden coffins, and a final inner coffin of pure gold, this casket alone weighing more than two hundred pounds. In fact, so devoted were they to this nesting of coffins that hidden among Tut's accoutrements was another miniature coffin, this one merely a vessel for a series of yet smaller coffins.

It's hard to say what makes nesting objects like these so pleasing, but you only need to recall your first encounter with them to remember the weird joy they bring. There is the simple surprise of discovering that one thing *contains* another, as well as the slightly obsessive-compulsive satisfaction that the object contained is merely a miniature of the container itself. But then comes the inkling, very dim at first, almost subconscious, that the thing contained might itself be a container. You open it, revealing—the mind boggles as before a precipice—yet another perfectly fitted object. With this the real business begins, the manic frenzy of opening ever smaller and more adorable containers that leads, if it goes on long enough, to a wholly unforeseen despondency: no matter how precious the final object, we will be sad to encounter it. It will be a disappointment for the simple reason that it cannot be opened. We do not want the process to end.

It reminds me of the relentlessness of the mind in grief, which can only be satisfied by exhaustion, or of the ever more nuanced disagreements of the Content Committee of the United States Holocaust Memorial Museum, or of the way an essay sometimes pursues its line of inquiry, finding inside one question yet another, and then another, and then another, each one a casket smaller and more intricate than the last. We open, open, open, and then just as we begin to feel dizzy, we arrive at a container, which upon being opened reveals, like the innermost coffin in Tutankhamun's tomb, a single lock of hair, plaited and auburn and smelling of some fragrant unguent, the hair perhaps of someone once loved. And then we come to a conclusion that is not a conclusion at all, a conclusion similar to the one reached by the Content Committee, which chose finally not to exhibit the hair but neither to return it to Auschwitz. It was sent instead to a storage facility outside of DC, and a space was left open in the exhibit hall, a place where the wall of hair could one day stand, when we are ready for the way it will make us feel.

XII

Tears

It was foul, and I loved it.

> Augustine of Hippo, *Confessions*, on the subject of crying

---

Terms

n. tear, teardrop

v. weep, sob, cry your eyes out, cry your heart out, start the waterworks, shed tears, whimper, snivel, howl, blubber, bawl

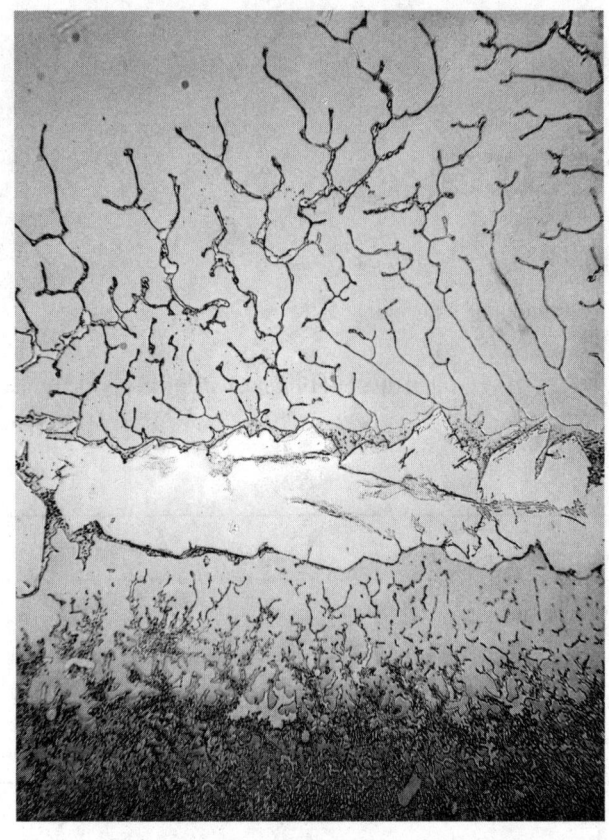

Crystallized tear, *The Topography of Tears*, Rose-Lynn Fisher, 2017

Biological Prologue: Where Tears Come From

Contrary to what many assume, tears don't come out of the tear ducts at the inner corner of the eye. The ducts, in fact, are siphons that drain tears *from* the eye into the nasal passages, which is why crying often makes the nose run. Tears are actually produced for the most part by glands hidden above the eye and released from way up under the eyelid. (You can glimpse part of the gland if you pull your upper eyelid up and out, though there's not much to see other than a veiny pink bulb.) But the belief that tears come from the ducts in the corner of the eye is a good example of just how little we understand about tears. Even in the most basic physical sense, most people don't know where tears come from.

When we think of tears, we tend to imagine them in a raindrop shape, but that's not really the form tears take most of the time, and when you think of them as a droplet, you're missing out on one of the main reasons tears exist. The purpose of tears is to keep the eye happy. Generally they're found not in a raindrop form but spread out over the surface of the eye in a layer a few micrometers thick, what's called the tear film. This film itself can actually be divided into three layers, each with its own distinct purposes. On the outermost surface is mainly meibum, a thin oily layer with a low melting point that allows it to remain liquid even in the tear film. Much as cooking oil floats on water, this meibum floats atop the other layers, reducing evaporation, intercepting contaminants, and giving the eye a smooth surface for light refraction. Beneath this is a watery layer that lubricates and provides nutrients to the eye while washing away foreign material. (Since the aqueous layer is filtered from the blood, it contains many of the same components as blood, making it an excellent and minimally invasive diagnostic tool.) The innermost layer, what covers the eye's surface, is largely made up of mucins. In addition to giving the watery layer something to stick to, these mucins protect against bacterial adhesion.

Of course, you're always making tears. Otherwise, the eye would dry out. But you're not always making the same *kinds* of tears. In fact, there are three distinct tear types in humans. For the most part, you make basal tears, the everyday sort of tear that lubricates and nourishes and protects the eye. But there are also reflex tears: when a gnat flies smack into your cornea, for instance, or when a child pokes you in the eyeball with a stick, the lacrimal glands exude voluminous watery tears to flush any offending matter from the eye. Generally, if you see an animal crying on this planet—an elephant at the zoo, for instance, or a second-place collie after Westminster—they're making basal or reflex tears, no matter what the sentimental among us believe. But humans are different. We also produce a third form of tear—emotional tears.

As for the question of where these tears come from, there aren't satisfying answers. Darwin thought they were simply a byproduct of intense muscle contractions around the eyes:

> I have myself felt, and have observed in other grown-up persons, that when tears are restrained with

difficulty, as in reading a pathetic story, it is almost impossible to prevent the various muscles, which with young children are brought into strong action during their screaming-fits, from slightly twitching or trembling.

He believed that when we despaired as children, the muscles around our eyes squeezed out tears, and having habituated the neural pathway, adults cried when they were unhappy. Darwin also suggested that if we were only to train ourselves as children to the habit of laughter, which can also produce tears, we might find that we cried "from gentle laughter, or a smile, or even a pleasing thought."

We know there's a cluster of neurons nestled into the brainstem, the superior salivatory nucleus, that works to control the body's salivation, and here, basically renting out a desk in the corner of saliva's office, is the tiny group of cells in charge of crying. To simplify descriptions of crying, this group of cells is sometimes called the lacrimal nucleus, but it's really such a puny cluster that it hardly qualifies even for the designation of nucleus. Sensory input from the surface of the eyes—sharp stick—follows a few pathways to this corner of the salivatory nucleus and is combined with information from the central autonomic network, the interconnected brain structures that oversee many of the body's involuntary functions. The body's decision is then passed back to the desk, and from there it wends its way back out to the lacrimal and accessory lacrimal glands. And we cry.

As for what exactly in the brain initiates emotional crying, much of what we know comes from examining damaged brains. Lesions in the periaqueductal gray reduce distress calls in squirrel monkeys, and removal of the amygdala causes rhesus monkeys not to vocalize when distressed. Meanwhile, damage to the corticobulbar tract and fronto-limbic circuit sometimes induces crying without emotion. On a finer level, crying may be related to a specific kind of neuron known as the spindle neuron. These neurons, which evolved independently in animals with larger brains—elephants, whales, apes—are especially thick. It's theorized they evolved as a way to speed information across large brains but were then repurposed for the processing of social and emotional information. This is all still very hypothetical, but what's clear is that those with familial dysautonomia suffer from a deficit of spindle neurons, and they are also unable to cry. Beyond this, we know very little.

Crying seems to change the balance of activity between the sympathetic and parasympathetic nervous systems. These two systems act in response to the presence or absence of stressful stimuli, with the sympathetic system revving us up and the parasympathetic revving us down. If we come across a bear in the woods, the sympathetic nervous system increases our blood pressure, heart rate, and respiratory rate. If the bear departs, it's the parasympathetic nervous system that gears the body's processes down, relaxing the muscles, slowing the breath, and focusing energy on digestion. The parasympathetic nervous system, in other words, is the part of you that chills out. There's slightly conflicting data here, but it seems that as a cry approaches, there's an increase in activity in the sympathetic system while the parasympathetic activity subsides. The moment tears begin to flow, however, sympathetic activity immediately falls, while parasympathetic activity rises and, importantly, remains elevated long after the crying. But it can be hard to say anything for certain about what happens after we cry. As one researcher noted, reviewing the hodgepodge of crying research, despite how universal crying is, we cannot yet even say when exactly a cry begins, let alone when it ends.

A Dog, A Horse, A Book, A Glass

At 40–15, in the third game of the second set of the 2017 Wimbledon Final, Roger Federer hits a serve out wide to the deuce side. Marin Čilić (pronounced *chill*-itch), the tallest man ever to play in the final, stutter steps to receive the ball, but Federer's spun it in such a way that it just drifts farther and farther off the court. Though Čilić's wingspan, including a racket-length on either side, pushes ten feet, even he has no chance of catching the serve. It whips past him, thunking against the padding at the back of the court: ace.

In the grand scheme of things, none of this matters. If you don't follow tennis, the terminology is all gobbledygook, and outside of standing along a highway, most people don't have much frame of reference for the sort of hellish fluorescent laser a 118-mph serve constitutes. But what happens next is important.

Without breaking stride, Čilić walks off the court and over to his seat. This is an odd-numbered game, so according to tennis arcana, the players switch sides. They are permitted to sit in their chairs for precisely ninety seconds. At this changeover, as in almost every other, Federer arranges a blanket over his legs, nibbles a single bite of banana, washes it down with a single mouthful of some grayish hydrating solution. The image of self-control, he stares ahead as blankly as a bodhisattva. Čilić, meanwhile, sits down, and as with any truly large person, the furniture puts his unusual height in context. "The big man" is what the commentators insist on calling him. He has to fold himself into the chair, back hunched, knees at an acute angle, and this heightens the effect of what happens next. Summoned from off-court, the tournament doctor and a physiotherapist appear beside Čilić. There's nothing immediately extraordinary about this. He's been looking tight on-court, and trainers get called out all the time to deal with minor physical ailments, a cramping hamstring or sore knee. Some players (Rafael Nadal comes to mind) even seem to take a Total War approach to the game, employing the trainer specifically in moments when Federer is hitting his Zen-like stride, as a way of unsettling his focus. But that doesn't seem to be the case here. Čilić towels the sweat from his brow as the trainer bends to consult him. He's trying to say something but seems to be having difficulty getting the words out. Then, instead of removing the towel, he presses it harder against his face, and then, though it takes everyone a moment to realize it, he begins to cry.

You can't make out his face at first—a number of officials are milling in and out of the shot—but you can tell everything you need to know from the way his chest is shuddering. Much as vomiting or defecation requires that we cede to the body control of its own autonomic processes, a human weeping loses a great deal of say-so over their organism. The forehead is drawn into lines, the cheeks go rigid with spasms, the eyes blink rapidly, the skin goes blotchy, the nose runs. The diaphragm is heavily involved as well, and there's no mistaking right now how Čilić's frame convulses with each hyperventilated breath.

Not only do we see him from one camera angle, then another, sitting in a chair in the center of the stadium, shoulders heaving, towel pressed against his face, but the weirdness of the whole thing is compounded by the voices of the three commentators, men charged with upholding sports broadcasting's most sacred covenant: at no moment can there be simple regard. In ordinary circumstances, they narrate every second they can, one describing on-screen action ("He takes the forehand crosscourt . . .") while the others fill in downtime with historical factoids, strategic analysis, and funny anecdotes. The Wimbledon Championship being for tennis fans a kind of World Cup in white collared shirts, the three men in the broadcast booth have happily occupied themselves thus far by pointing out the famous people in the crowd ("Longtime tennis fan Dame Anna Wintour is here today . . ."). This, though, is an entirely different matter.

It will help explain the deeply anomalous nature of Čilić's tears that these men, some of the most naturally gifted chatterboxes in tennis, obviously have no clue what to say now. They have at their fingertips every conceivable tennis statistic. They can tell you the top speed of Marin Čilić's serve or the number of unforced errors he hit in the first set. They can tell you what Marin usually eats for breakfast and how Marin met his fiancée. With the assistance of IBM's Hawk-Eye Camera system, they could even pinpoint within millimeters the invisible oblate skid mark left by a 90-mph forehand on the court's faded grass. But they've neglected to brush up on Fox, Lapate, Shackman, and Davidson's *The Nature of Emotion*. Their copies of Sontag's *Regarding the Pain of Others* have been left back at the hotel. They are out of their depth. Exactly what sort of broadcast material is another human's sorrow? Is this play-by-play territory, or should they provide context and analysis? What is the strategy of a man weeping? They can't tell you. They stutter, they mumble, they trail off. For entire seconds, eons in the timescale of televised sport, there is dead air.

It's supremely uncomfortable. People are standing in the aisles, craning their necks to get a closer look. A camera swivels to Čilić's box, where his coach and agent sit side by side, the former sucking his teeth, the latter staring vacantly into space, as if watching a Nike contract slip softly down the shredder. You can tell by the way Čilić keeps putting the towel over his face how much he would like to be anywhere but here, anytime but now. Eventually, some instinctive urge to babble does kick in on the part of the commentators. They recall that crying happens all the time in tennis. Plenty of people have cried on-court. Juan Martín del Potro, the gentle Argentine giant, often hugs fans after matches with tears streaming down his cheeks, and Roger, such a student of the game, he cries all the time. There was that Australian Open Final just a few years ago, and that time after yet another loss to Rafael Nadal when he huskily whispered, "Oh God, it's killing me." Though always *after* a big win or a loss, the commentators remind themselves, the crying always comes after, never *during* a match, especially never during *this* match, the Wimbledon Final, the pinnacle of a sport that more than anything is supposed to embody a certain stiffness of the upper lip.

"Bizarre," one of the men whispers to no one in particular.

One man brings up the third set of the 2004 French Open Final between Gastón Gaudio and Guillermo Coria. In that match, the fleet-footed twenty-two-year-old Coria, up two sets with the game in hand, suddenly lost his ability to move or serve. Trainers came out to treat what were reported as debilitating leg cramps, but the issue, to many, wasn't in Coria's legs. It was in his head. The match is well known to have left not only Coria's career in tatters (plagued by the yips, he retired in 2009, aged twenty-seven) but also somehow Gaudio's. Among some tennis players, it has an almost Beetlejuician reputation, as if the very mention could unleash a psychological tennis catastrophe. The commentators are happy to brush past it quickly. Coria never actually cried, anyway.

Eventually a camera gets a good angle on things, and there's a close-up of Čilić's face, eyes red, mouth drawn, and there's no longer any way to talk around things: Marin Čilić, this dour, bearded, six-foot-six Croatian, is weeping on Centre Court and no one knows why. The second commentator asks if they've ever seen anything like this, and silence blooms again on the broadcast.

The reply, when the third speaks, sounds dry on the palate; "Not in this type of situation, no."

ONE DOESN'T JUST CRY. ONE GETS CHOKED UP. ONE IS moved, or driven, or sometimes reduced to tears, one's eyes go glassy, one wrestles with, struggles through, fights back tears, and then all at once, one tears up, tears spring to one's eyes, one bursts into tears, tears stream down one's face or streak one's cheeks, and one's mascara runs. One groans, heaves, grieves, wails, laments. Occasionally, one melts, weeping over one's misfortune, and is wracked by sobs. One's cup runs over with tears. One cries buckets, rivers, sometimes seas. One drinks or feasts or gorges upon one's tears. One never just cries. One cries one's heart out, like a baby, and then, only then, as Homer says of Achilles, can one dry one's eyes, having had one's fill of tears.

While the processes of the body merit careful attention and description, being so central to the experience of being human, the secretion of fluid from a gland rarely enjoys such profuse phraseology. It's hard to say exactly why tears demand it. Possibly, it's a testament to the essential ineffability of the feeling of crying, how in weeping we seem to touch the shores of some wild and illiterate continent. Between the mind that's crying and the one that isn't, there seems to exist some gulf that language cannot cross. It's almost as if the neural circuits that give rise to tears cannot coexist with those that facilitate language. We can attempt to describe someone crying or to put words to our own experience of crying, but the logic of language only carries us further from the actual feeling, dooming us to endless euphemization.

Though humans are the only animals that physically produce emotional tears, many infant mammals, when in need of attention from their caregivers, make unique sequences of sounds known as distress calls. Parents are often able to unerringly differentiate the calls of their own specific offspring from those of others'—as can be seen any day of the week when some near-indistinguishable cry on a raucous playground causes precisely one parent to suddenly look up—but distress calls are in fact so sonically similar across different species that it's not uncommon for people to mistake the mewling of a kitten for the crying of a human child. The calls are so universal that they seem to be of ancient origin in the mammal family tree. They're theorized to have evolved, in part, in response to the mammalian innovation of continuing gestation outside the body, with selection favoring those parent-offspring pairs whose brains developed coordinated automatic responses.

In humans, so the theory goes, this capacity developed still further, outlasting infancy and taking on a more general prosocial role of reinforcing bonds among members of the species, while also evolving, for confounding reasons, its signature overlubrication of the eyes. So it could also be that a gulf always exists between any two minds. Language helps us knit that gap. Crying is merely a more ancient, and perhaps more fundamental, way of getting two animals on the same page. In this respect, tears would always be a kind of invitation. I feel that. Of all the stimuli that cause people to cry, one of the most potent is seeing someone else cry. That could be why Čilić's tears feel so uncomfortable. Tears ask for company.

There's some evidence that tears played precisely this larger social role in our past. In the first extant mention of tears—a fourteenth-century BC clay tablet from the lost city of Ugarit—the goddess Anat, finding the corpse of her brother, Ba'al, throws herself across his body, weeping. "She drinks her tears," that's what the text actually says, "she drinks her tears like wine," and some scholars believe the story in fact formed the basis of an archaic Canaanite planting ritual, with analogs in the traditions of Egypt and Mesopotamia, where each spring, having journeyed together outside their settlement, an entire tribe, playing the role of Anat, would begin to whimper, then wail, the feeling gathering in intensity over several days until, transported into an almost trancelike state of hysterical weeping, intoxicated by tears, they collapsed as one into fits of laughter, giggles, and other expressions of irrepressible mirth. Then they'd plant their fields.

The behavior sounds aberrant now. It's hard to imagine getting together with a bunch of friends to cry and grow some corn. But when you see the uniform look of anguish on the painted faces of half a stadium as the home team loses, when you hear the sniffles erupting from across a movie theater, when you think how 9/11 remains for many a touchstone, a rare moment of national and global emotional unity, the idea of crying as a social bonding practice doesn't seem at all far-fetched. Crying, after all, is a peculiarly intimate act. It's tied, in unclear ways, to neuropeptides like oxytocin, vasopressin, and prolactin, which are all recognized for their role in inculcating feelings of social attachment in animals. I've wept while reading the Gettysburg Address to my daughter at the Lincoln Memorial, for instance, but I've also cried during the second *Jumanji* and at the end of Magda Szabó's *The Door*, an obscure 1980s Hungarian novel about a militant housekeeper. The three things have little in common except for my feeling an almost inexplicable devotion to that housekeeper, to Abraham Lincoln, to my daughter, and to Dwayne "The Rock" Johnson.

Though it's relatively easy to recognize crying, we still don't know much about where tears come from, the "neural structures underlying cry production," as the scientific literature prefers it. We don't even know, neurologically speaking, what exactly emotions are, even though they act as almost physical transformations of the brain, changing the information we have access to and the way we interpret it. Emotions seem to be transient patterns of neuronal activity that vary in duration, intensity, parameters, and degree of self-reinforcement and that are subject to modification by external circumstances (social context, cloud cover, heights) and internal circumstances (low blood sugar, slow breathing), as well as by the activity of the brain itself (PTSD, meditation, habit), which is to say they're not only wildly variable and unpredictable but also inextricable from a context that includes the entire conscious and subconscious memory of the crier. They're difficult to study, in other words, and it's hard to say, given any specific cry, where exactly the tears are coming from. People end up crying, like Čilić, in all sorts of unexpected situations. Some cry in the crush of tourists at Notre Dame. Many, according to a survey by Virgin Atlantic, weep on international flights. In Jerusalem they cry before a retaining wall, and in Graceland they sob in Elvis Presley's mirrored basement. Tear-strewn faces perpetually crowd the two big holes in the ground where the Twin Towers once stood. New York itself, maybe due to the height of its buildings, maybe merely its density, is a mecca for tears, and on any given day, it's not uncommon to see someone crying on a park bench or street corner. Something in the metropolis just wants to cry. "New York City," as William Maxwell wrote, "is a place where one can weep on the sidewalk in perfect privacy."

Though tears seem to have evolved as a social behavior, privacy seems paramount to any understanding of them. Crying is often accompanied by feelings of vulnerability, and much as Čilić seems to, many find this sense of vulnerability acutely unpleasant. They are only comfortable crying in private. Some cry in closets or confessionals, others in the dark anonymity of the movie theater. The design of the car, its sense of insulation from the exterior world, makes it particularly conducive to tears, and in a business book I can no longer find, I remember reading about a traveling milkshake mixer salesman who began to weep with such frequency and ferocity during his long drives across the Midwest that he was forced to abandon his job, opening instead, in 1955 in Des Plaines, Illinois, a roadside hamburger restaurant named McDonald's.

For some, crying seems to require not only privacy but also familiarity. These people can experience all manner of suffering in their lives, but only at home do the floodgates open. The threshold exacts a sort of Pavlovian response from their organism, the homeostatic functions changing in reaction to a familiar suite of sensory stimuli so that the entire internal state of the body shifts gears: metabolism, respiratory rhythms, hormone secretion, glucose tolerance, fat deposition. Just as these changes often manifest themselves in a loosening of the bowels, they can also result in tears. "While I sit at home sometimes hot tears come," as Menelaus says, recalling the Trojan War, "and I revel in them." Some actually seem unable to cry outside the home. Feeling emotion surge through him in the delivery room during the birth of his first child, an outdoorsy lawyer friend of mine simply passed out. It's common apparently. As an obstetric nurse once told me, pressing me into a chair, "I like to make sure Dad is sitting down."

Some carry things a step further, finding the urge to cry rising in their chest only once the bathroom door has clicked shut behind them. I get it. Not only is the bathroom the primary site of our bodily disintegration, it is also often, in the most basic physical sense, the house's safest place: compact, lockable, often located at the rear or interior of a home with windows that, if they exist at all, are usually few and small. It's no accident those seeking refuge from tornadoes are counseled to take refuge in their bathtubs. In the modern bathroom, if anecdotal evidence is to be believed, there is a still more private space, so that while some cry on the toilet, others before the mirror, many will cry only in the shower, as if in sympathy with the showerhead. In fact, the shower is said to host our tears with such frequency, and the tears cried there to be so irretrievably abject, that the

English language has evolved a mildly pejorative verb for the act of weeping there: to shower-cry.

BY THIS POINT, I'VE CALLED MY WIFE IN FROM THE other room. She avoids tennis, especially the Wimbledon championship, with its promise of several hours squandered indoors on an otherwise gorgeous Sunday in July, but she will on occasion briefly humor me. "Why isn't anyone going over to help him," she says as soon as she walks into the room, and of course she's right. Anyone can see how desperately alone he looks out there. Here's a man crying while fifteen thousand people mill around, and no one is doing anything about it. The referee is sitting above Čilić in his green lifeguard's chair, occupying himself a little too studiously with something on a tablet computer. Federer is just on the other side of him, staring off into space with his blanket folded just-so over his knees.

You find yourself sort of secularly praying at Čilić to pull himself together, to feel better, *we love you get up*, which of course doesn't do any good, so instead you feel angry at all those people who aren't doing anything, at the referee and Federer and the commentators and fans. Though what can they do? While seemingly designed to eventually precipitate it, tennis isn't prepared for this moment. Čilić's coaches are specifically barred from interacting with him during a match. The security detail isn't just going to allow a consoling fan onto the court, and it's not Federer's place to go over and get up in Čilić's business, especially since he is implicated in all of this in a way that he would obviously prefer to ignore, at least for the time being and probably forever. The trainer puts a hand on Čilić's shoulder, but it's not the full-hearted gesture of sympathy the moment requires.

Though tennis is widely recognized as a uniquely psychological game—as evidenced by the popularity of books like *The Inner Game of Tennis*, *A Champion's Mind*, *Mental Tennis*—there seems to be a resistance on the broadcast to the idea that whatever Čilić is experiencing out there is primarily mental. One of the men, a former player renowned for channeling his on-court feelings into torrents of umpire-directed invective, is uncomfortable even acknowledging that the tears are emotional at all. There must be some kind of injury, he insists.

"We haven't seen any kind of treatment," says the second. "We've seen . . . tears . . ."

"Did he hurt himself when he fell?" the first says, referencing a minor slip earlier in the match. "Did . . . I . . . I don't see any other possible explanation . . . he got hurt . . . something's hurting . . ."

"There was no attempt to treat anything with the body," the second man reiterates.

"Oh boy," the first murmurs. He draws away from the mic as he says it, lending the words a feeling of tenderness and disorientation. The comment seems to echo out to viewers from across an empty plain.

IT'S HARD TO KNOW WHAT TO DO WITH YOURSELF IN this situation. The commentators seem to be modern, sensitive men. They recognize the realities of stress and anxiety, the need for empathy, work-life balance, a positive mental attitude. Earlier in the broadcast, they were even approvingly discussing the way Čilić's new coach has been working with him on expressing emotion rather than bottling it up. And yet none of them is willing to mention on air a time he actually cried.

People generally, and maybe men especially, have long been suspicious of tears. Tears resist description, they've often been associated with women and children, and to many, they give the appearance of weakness, maybe even unreliability. No one, glancing into the cockpit, wants to see the pilot crying. But it's more than that. We're reasoning creatures. Thanks to little more than a pale mush of a hundred billion neurons, we've largely subdued a planet and puzzled out much of the inner workings of a cosmos. We like to think, and whether we adopt the title or not, each of us is, in their own way, a philosopher. We sift through incoming data, generalize, abstract, synthesize, deduce, resolve contradictions, and in so doing, we slowly assemble a consistent, cohesive model of the world, governed by fixed rules. Much like mathematicians, given x, we learn to expect y and z. Emotion, and maybe especially the extreme emotion of crying, threatens this project at its very foundations. One and one can't suddenly make zero when our mother dies. Infinity doesn't get larger because we skip breakfast. But these are the sorts of effects emotion causes, physically changing the way the brain operates so that the same inputs result in different outputs. Reason seems to play an uncomfortably subordinate role to the emotions.

It's not surprising then that philosophers in vastly different traditions have attacked tears. Socrates scolds his followers for crying at his deathbed like women, and Zhuangzi, gaily singing and drumming after his wife's death, maintains that, since our existence may be a dream, we should not cling to it. When we cry, he tells us, we are being uncomprehending toward destiny. Seneca says we own nothing, not the food we eat, not the tennis racket we hold, not the record for fastest serve, not even our lives. We are going to die, and so is everyone we know and love, and tears only show that we haven't accepted this most essential fact. To Seneca, there really is no appropriate time or place to cry. "What need is there to cry over parts of life," as he writes to a grieving mother. "The whole of it calls for tears." And Montaigne, debating in an early essay which was the greater, Democritus, the laughing philosopher, or Heraclitus, the weeping one, sides decisively with the former: "Not because it is pleasanter to laugh than to weep, but because it is more disdainful, and condemns us more than the other." The problem with tears, he says, is that they find value in existence.

But Montaigne is an interesting case. A wealthy landowner in Périgord, he is immobilized in 1571 by a fear of death. Rather than sinking into the distractions of administering his estate, however, he determines instead to throw himself into a life of contemplation. Leaving behind many of the rights and responsibilities of his position, he installs himself on the uppermost floor of a medieval tower with little more than the books in his library as company. The ultimate object of philosophy is to learn to die, he writes. He hopes that by contemplating his own demise, as Seneca counsels, he will dispel its terrible mystery.

He intends to live a life of the mind, in conversation with the ancients, and he is inclined, at first, to the somewhat depriving approach of the Stoics. He looks with disdain at fine foods and other sensual pleasures. He considers attachment to earthly possessions folly and regards tears with suspicion. But then something strange happens. He develops kidney stones. "The obstinacy of my kidney stones," he writes, "especially in the penis, has sometimes cast me into long retentions of urine, for three, even four days, and so far forward into death that it would have been madness to hope, or even to wish, to avoid it." The affliction overwhelms his sedate existence, sending him abroad, in vain, to seek treatment and forcing him into long periods of suffering and convalescence. But as he labors to evacuate the small stones that will eventually kill him, he continues following the relentless train of thought that defines his writing, throwing his mind into "subjects as remote as possible from my condition" and finding, unsurprisingly given his state of duress, a different perspective. Abject, crying out in pain, he begins to realize the limits of philosophy, writing in one essay that no matter how great a man's wisdom, it cannot grasp suffering so well as eyes and ears. He doesn't abandon philosophy completely, but he recognizes, ultimately, that it can only bring him so far. What philosophy formed, he writes, the stone perfected. He has been within a hair of death. He knows he is bound to life by only the most slight and frivolous of things: a dog, a horse, a book, a glass. He is no longer afraid to die.

And he breaks with the Stoic attitude toward tears, finding, he says, too formalistic the Stoic precept that so rigorously and precisely orders us to maintain a good countenance and a disdainful and composed bearing in our suffering. Let philosophy be content with our inner workings, he says, and leave to actors the care of external appearances. We may be steadfast in our heart and yet grant pain its voice.

SOME SEEM BETTER AT CRYING THAN OTHERS. THEY seem to experience the world with an exponentially higher fidelity. They're able to cry anywhere, weeping in dining rooms and office corridors, in elevators and on crowded streets. "Who can say what it means to be sensitive," as William Maxwell writes, "considering all there is to be sensitive to." There's Ryōkan, the eighteenth-century hermit-monk who is brought to tears by the sight of children returning at dusk from their games, each to their own home, as their parents call their names for dinner. "My sleeve is wet with tears" is how more than one haiku end. The mere sound of leaves falling outside his hermitage makes him weep. Not only does Augustine weep and weep, but detailing the perverse enthusiasm he feels for crying, he sounds as if he's describing an addiction. Francis goes blind with tears, and Nietzsche, one day, seeing a draft horse being beaten in the reins, falls into such a shambolic state of weeping, his arms thrown around the animal's neck, that his mind never recovers. I think of Abraham Lincoln most of all, of the ease and violence with which tears found him, how he'd often pull down

Shakespeare or the Bible, as if to hone with verse the feeling's edge. One day in the White House while composing a speech, he is interrupted by his friend Orville Browning. He looks up, his long, gaunt face streaked with tears. "Browning," he says hoarsely, struggling to gain control of his voice. "We are all going to die."

He's right, of course. Each passing second draws us away from everyone and everything we love. That's what puts the extra beat in the kiss good night. It's what draws the skin tight around the eyes of the father of the bride. It's why the hand hesitates before putting the car in park for the last visit to the veterinarian. It's the dog and the horse and the book and the glass, and I suspect it's part of what's going on behind Marin Čilić's towel on Centre Court. We're going to die. It's no great metaphysical statement. It's crying. Baby mammals do it all the time.

It hasn't escaped my attention that the composition of this book has coincided almost exactly with the entrance into my life of a small and at first wholly helpless human being, or that my time over the past years has been dominated by the task of guiding this child into general continence. I have cleaned her excrement off blankets. I have sung her songs as she sat astride a miniature toilet, her eyes closed with focus. I have mopped her vomit, cut her hair, bandaged her knees, warmed her milk, accepted in my hand more than once the benignant offer of a booger. And still, every so often, I wake in the dark and can't fall back asleep until I've gone to her door and listened for the sound of breathing.

But it has all been nothing, less than nothing, compared to

the work of accompanying her tears. Calm is like a little husk she clutches feebly to her shoulders. The least thing can blow it away, plunging her into an apoplexy of grief so wretched and base that for a full ten seconds she can only draw in breath. The fact that we don't know the neural structures underlying cry production strikes me as such a foundational lack, you see, because at some point, at many points, almost daily in fact, I have to consider, like those men in the broadcasting booth, whether this is the time or place for tears. I have to help her to stop crying. We call it creating emotional resilience, but in many ways I'm just teaching her how to put up a front, how to drive her thoughts toward less morbid pastures, how to feel less. It's a tough lesson to get stuck teaching. Why shouldn't we cry all the time? The more I love her, the less I know.

MARIN ČILIĆ WILL GO ON TO LOSE THE FINAL, A LOSS that will, in many ways, mark the culmination of his tennis career. In a charming attempt at locating his pain someplace simple, physical, he'll attribute his tears in a postmatch press conference to a number of severe blisters on his feet. Though his time on the court will not have been without rewards—millions in winnings and endorsements, a championship at the U.S. Open—he will never achieve the seemingly immortal fame of the opponents who so often bested him: Federer, Nadal, Djokovic. A few months later, his ranking will peak at number three—he will be, for twelve weeks in the spring of 2018, men's tennis's third best—but he'll never have his name engraved on

Wimbledon's gilt cup. He'll never even return to a final Sunday on Centre Court. Instead, he'll get older, as all tennis players do, his power and his form degrading imperceptibly, little by little, day by day, his injuries accumulating one by one, each recovery taking longer than the last, and he'll recede softly from a spotlight no one ever really owns, becoming a footnote, a heel, an also-ran, until at last he slips down the rankings, as all tennis players do, as we all do, like a stone.

But not yet. Right now, that's but a premonition. Right now, he's still on Centre Court, still crying, and the commentators are still trying to figure out what to say about it. One of the men recalls that a certain tennis truism might apply here—tennis is the loneliest sport, etc.—and he trots this out only to watch it wither on arrival. It hasn't even been a minute and a half, but there's some obvious desperation on the telecast, maybe even depression. The big man has glimpsed something you're not supposed to glimpse while playing tennis, and now we've all glimpsed it too. You can tell the commentators would like to just go back to talking about second serve percentages and who on tour has the softest hands at net, but the path back there from here isn't exactly clear. It almost feels like these three men could use a good cry. Perhaps we all could. I find myself wishing the commentators would do what Mr. Rogers once did at an awards ceremony, plunging a room into tears merely by inviting those present to join him in taking ten seconds of silence to remember the people that loved them into being. "Ten seconds," he said, looking at his watch. "I'll count."

You won't find Čilić's tears on the official highlights, of

course. In fact, the highlights don't even show those first three games of the second set, as if the tears retroactively tainted the moments that preceded them. Instead, the reel cuts directly from the end of the first set to the beginning of the fourth game in the second, and you completely miss what made that fourth game so challenging to spectate upon, so emotionally discombobulating, that for a long time afterward I could not watch tennis at all. Because there is a moment toward the end of Čilić's episode (*episode* feels like a paltry word for it, but one paltry word is as good as any other here) when the chair umpire calls time, Federer walks to his side of the court, and the physiotherapist and the doctor and the tournament referee suddenly evaporate. All at once, Čilić is more alone than ever. He seems to sense it. He reflexively throws the towel over his head and starts pouring water from a bottle into his hand and splashing it up onto his face. It's this moment: when he must wipe away his tears, stand up, walk to the service line, and *continue playing tennis*. In a dim way, we all knew it was coming, but its reality is far worse than imagined. The crowd cheers for him, but he can't hear it, and if we felt guilty watching Čilić cry, now we just feel aimless and confused. The whole atmosphere is deflated, and the knocking of a little felt ball over a waist-high net feels exactly as arbitrary as it sounds. Čilić bounces the ball a few times, shoulders slumped. Even Federer, so eminently unflappable, seems to have lost his normal cat-watching-mouse attention to the movements of the ball.

As Čilić sends a volley long, one of the commentators poses a final question to his colleagues in the booth.

"How in the world," he says, "do you compose yourself after what he just experienced?"

This question, I suggest, is more applicable to our lives than we might prefer to believe.

The peculiar thing about the ancient rituals of crying, after all, is that they were not rites of despair, so far as we are able to understand, but of rebirth. They were springtime festivals. Tears, to the ancients, were endowed with the power of resurrection. Ba'al, clasped in the arms of Anat, his face wetted with her tears, opens his eyes. Osiris awakens to a Nile flooded by Isis's grief. The soil, watered with sorrow, is prepared to bear life once more. It makes sense. Along with their social role, tears also seem to regulate the emotions, increasing the activity of the parasympathetic nervous system, leaving the body and mind in a calmer, clearer state. We may not always feel better after crying, but we rarely feel the same. Some describe a sense of refreshment or awakening. And maybe it isn't so much a question then of where tears come from but where, if we let them, they can take us.

Stricken by the death of his beloved Annaeus Serenus, Seneca leaves Rome in search of the meaning of his tears. Before he drums, Zhuangzi first weeps over the body of his wife, and it is in weeping that he glimpses the incomprehensibility of destiny. And Montaigne begins his *Essais* only after the death of Étienne de La Boétie, the man he loved most on Earth. He's not often mentioned, La Boétie, but you can see him lingering there in the margins of each page, this great friend. It's his presence that lends the work, even in its most difficult moments, such joy. And I'm not surprised that the rituals of Anat, those festivals of outdoor crying, should have survived to the present day, albeit in altered form, absorbed into the religion of the immigrants who settled in Canaan, and passed along throughout the generations, so that readers are still reminded today by the Book of Psalms to sow their fields with tears.

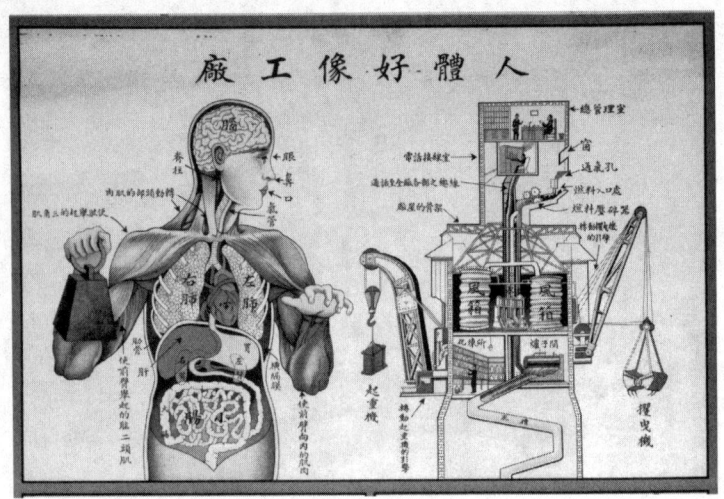

Chinese health education poster, 1933

Epilogue

The Body

> It is as if the organism itself were only an excrescence.
>
> Henri Bergson, *Creative Evolution*

Assuming half a liter a day by volume and a lifetime in the vicinity of seventy-five years, you'll leave behind about fourteen thousand liters of feces when you die. The sum of your urine will be closer to thirty-three thousand liters, almost exactly the size of a standard shipping container and only slightly less than the accumulated volume of your flatulence. You'll have made about six thousand liters of tears, though even for the most emotive of individuals, the portion derived from feelings will make up a minuscule fraction of that number. You might have produced a few thousand liters of milk, meanwhile, depending on your approach to reproduction, and you might have donated a hundred and fifty liters of blood, depending on your feelings about needles and markets. For all the hullaballoo surrounding ejaculation, the total semen production of even the most alacritous masturbator could be contained handily by a shelf of two-liter soda bottles, and though a period sometimes seems as though it will never end, you could not paint a house with the forty or so liters of menses produced during your lifetime. You'll have made a great deal of mucus, though, close to a hundred thousand liters, and an amount of breath, about three hundred million liters, that cannot easily be conceptualized. When Atropos snips the thread of your life, your head hair, measured as a single strand, will stretch close to two million feet.

All of these numbers are obviously rough estimates. You could live to 108 and weep an extra liter of tears or die tomorrow of dysentery. We should not, as Seneca reminds us, presume anything but that these breaths are our last. These materials are fundamental to who you are and how you get along with other people, but when they're quantified like this, I find the physical totals a little less than impressive. That two-million-foot hair would hardly get you from Providence to Baltimore, and you'd need a hundred lifetimes to fill a single Olympic swimming pool with urine, five or six hundred to fill it with tears. Your three hundred million liters of collective breath, though seemingly a staggering volume, represents such a puny fragment of the atmosphere that you'd need to chase the number deep into a field of decimals to distinguish it from zero. This is what you will leave behind. Not much.

The reader, by now, will have realized that this is a book about death. It comes as a surprise to me, as well. Though I suppose it shouldn't be entirely unexpected. Every movement of the bladder or the bowels, every period, every fallen eyelash, every exhalation, is a literal disintegration. Little by little, we are falling apart. This is part of the discomfort with discussing our bodily materials, isn't it? Each turd, each tear, is merely an interest payment on a principal that must eventually come due.

Of course, you'll leave behind another thing: your body itself. In some ways, this is shed at the end of a life too. I don't mean that in some spiritual sense of the soul shedding its earthly trappings. I only mean that eventually the body is just another thing that gets returned to the larger world. It's uncomfortable to think of your body in this way, in the same category as feces and hair, but that's the way it is. Despite the desires of countless theologians, the trajectory of your body's final journey will be less like the fiery passages of the stars and more akin to those meandering pilgrimages taken by your feces and your urine, your blood and vomit and tears. It will become something that must be dealt with, something that must be disposed of. We may disagree over the existence and nature of an afterlife, but everyone recognizes the stench of rotting flesh.

Perhaps nothing communicates the body's ultimate dissolution so much as the great pains we have taken to disguise it. Depending on where and when you die, you might be mummified à la Tutankhamun, your organs removed and replaced with fragrant spices and salt, your skin plastered with old receipts. Or as is still practiced by the Toraja of Indonesia, you might be preserved and entombed and brought out once every few years for drying and family photographs. You might be left out in the open air to live on in the form of the vultures that disembowel you, or like the Tollund Man, you could be submitted to a peat bog upon your death, so that your bones dissolve in the acidic environment while your skin remains perfectly preserved down to its final expression. Should you die in the United States today, your body might be burned in an oven and the resulting ashes stored in a decorative urn, though these ashes are in fact not ashes at all but tiny fragments of bone. Or a mortician might drain your blood and flush your circulatory system with a formaldehyde solution tinted with dye so that rather than appearing to your loved ones with the pallor of death, you greet them for the final time looking as though you've just returned from Palm Springs. Or like the Romans who inspired Sir Thomas Browne's *Urne-Buriall*, you might be crammed into a clay pot with your hairbrushes and rings and buried on the frontier. Should you experience a bodily resurrection, it will come not in the form of an angel but of a backhoe, as occurred recently in Des Moines, when several ancient bodies were unearthed during excavations for a new sewer system. Who knows the fate of his bones, as Browne says, or how often he is to be buried.

All this pomp and preservation, of course, only obscure the reality of what is happening. The body is becoming waste, and no matter how large an obelisk you plant atop it, the amount of matter that's actually left at the end of a human life is still quite paltry. Anyone who's held a shoe box full of ashes knows as much. If the eight billion people on this planet—roughly half a cubic kilometer of brains, guts, blood, bones, and mucus—were spread evenly across the terrestrial portion of Earth, the

EPILOGUE

resulting layer would be about as thick as the film of tears on your eye right now.

It's depressing. You can see why the old theologians spent so much time debating the ins and outs of bodily resurrection. Decomposition has a way of making you turn your thought in other directions.

What, if anything, remains then? In the most purely physical sense, your body contains about five hundred megajoules of energy, enough to run a lightbulb for a year or to drive a midsize sedan a hundred miles, or, to put things in the more familiar notation of digestion, roughly 120,000 kilocalories, the equivalent of a hundred Big Mac meals. This energy, stored in the form of chemical bonds, namely as molecules of glucose, protein, and fatty acids, will remain after you die. It only needs to be converted into adenosine triphosphate to continue its chemical journey in the shape of another. Since no creature will be capable of digesting your body in its entirety, the scavenging of this energy will take the form of a vast buffet. The glucose in your thigh muscle might be catabolized via glycolysis by a rat while the slender threads of a keratinophilic fungus hydrolyze the proteins in your hair. The real prize at this feast, however, will be those molecules that most efficiently store energy, your fatty acids, so that the caloric orgy reaches its apotheosis—it seems fitting—in that fattiest of all your organs, that thing which seemed most *you*, your brain.

You'll leave behind more than energy, though. If you've had children, your DNA will remain in circulation, as well. It will be an adulterated version of your DNA, and the proportion that's in any way you will rapidly dwindle with each passing generation. (Your grandchild's grandchild will be 94 percent other people, just as you are only faintly related to your grandmother's grandmother.) But it's something. You'll survive like this, in a shade of hair, a loop of the intestine, or a disorder of the blood, much as you yourself are a patchwork of your ancestors. Even if you haven't had children, you'll have propagated and spread a great deal of genetic code simply through the viruses and bacteria in your body. None of this is negligible. A child can create a civilization, just as a single virus, sheltered and evolving within the human body, can destroy it.

You're bound at this point too, though, to start considering what really qualifies as you. Has it all been about chemical energy and ribbons of genetic data? Are you merely a byproduct of their mute Sisyphean need to replicate? Lips rouged, hair combed, your body gets dumped into a hole while the genes go on forever copying themselves like some mad admin chained to its Xerox. Is that the real story?

There is another possibility. Though not material in the same way as urine or tears, our bodies do emit one other thing: sound. It is an ephemeral effluvium, little more in our species than a perturbed exhalation, and though we might be tempted to accord it a different, perhaps a higher, status from that of our feces and our hair, it is no less of an emission. Our lives are saturated so completely by this particular waste—from the TV commentator's joke to the lover's whisper—that it is difficult to think of it in the same category as the bacteria-laden effluent we flush down the toilet each morning, but our first sounds, far back

EPILOGUE

at the base of the evolutionary tree, would have been precisely this: nothing more than a byproduct of the body's functioning, like urine, only in the form not of matter but of energy. Much as milk, in its first incarnation, was merely an excretion, and much as the tears of mice once held no meaning, so did our first sounds echo over the Earth, signifying nothing.

The study of the evolution of sound is so constrained by a lack of information as to be almost laughably theoretical. *Homo erectus* had no reel-to-reel, and the organs with which we produce sounds—the tongue, the lips, the throat and larynx—are composed entirely of soft tissue and cartilage. They leave no fossil record. What is quite clear, though, based on the prevalence of ears today, is the profound and near-universal advantage conferred by the ability to hear. Not only did auditory organs evolve independently across lineages, but they also diversified so widely within them that the world is now filled with countless variations on the theme. Squid, for example, hear through an inverted tennis ball–like organ at the base of their brains, while fish hear via denser-than-water calcium carbonate structures called otoliths. Sharks use crystals known as otoconia. In insects, hearing evolved no fewer than nineteen times from the proprioceptive stretch sensors located between the segments of their bodies, which is why cicadas, crickets, and fruit flies hear from their abdomens, legs, and antennae, respectively, while the dainty lacewing hears with its wings. On our own branch of the evolutionary tree, the ear as we would recognize it today evolved during the Triassic as our newly terrestrial ancestors adjusted to the novel acoustic medium of air. The result: the independent evolution, in snakes, turtles, crocodiles, birds, and mammals, of the tympanic middle ear, where sound waves vibrating a membrane are passed through the stapes bone into a snail-shaped fluid-filled chamber, thus transforming sound energy into the fluid motion by which hair cells send information to the brain.

As this book has shown, our wastes often—I almost want to say inevitably—play a social role. Certain crustaceans in the Caribbean find mates by sneezing out a luminescent mucus, hippopotami culture revolves around the dung midden, the common dog glues its nose to the telephone pole for news of its kind, and with the power of the ear to locate and assess mates and offspring, sound too becomes part of the social fabric, from the singing of whales in the Pacific Ocean to the distress calls of mice in the lab. But in one species, this waste takes on a remarkably outsize role. So fundamental is it, in fact, that the evolution of the creature can hardly be understood without regard to the manipulations it makes of sound.

It is an ape, and at some point in its evolution, its tympanic middle ear borrows two bones from the jaw, a more complex arrangement that may give the structure a higher degree of evolvability. Little by little, as the ages pass, its neck lengthens, shedding the air sacs and vocal folds, so distinctive in other primates. Its larynx descends. Its tongue and lips become more agile. In the lungs, the nervous system extends its domain, dramatically increasing innervation, and in the brain, the auditory cortex develops into a complex network of interconnected regions. The result is that, unlike other primates, this creature, *Homo sapiens*, is able to create precise pitches with its vocal cords, to sustain them with steady air pressure from the lungs, and to manipulate them with rapid, coordinated movements of the throat, tongue, jaw, and lips, creating dozens of distinct sounds in rapid succession. Not only this, but with its ears, it can localize and isolate a sound source, filter it from background noise based on its acoustic features such as pitch and timbre, and differentiate, in real time, an endless slur of sounds into syllables, words, sentences. And though we can't know with any certainty, at some point during this process, probably at many points, slowly, over echolalic eons that dwarf our capacity for narrative, this creature makes the leap from purely expressive sound—the grunts, whinnies, and squeaks of its fellow animals—to symbolic speech. Language is born.

It's strange, but this moment, when sounds become descriptive of more than just the body, this is when the waste differentiates itself from the others. It becomes a code, and much like DNA, it is able to survive the destruction of the material that hosts it, a catchy tune or turn of phrase propagating itself through a family, a clan, a culture, perhaps even taking on the shapes of stories. It goes on like this for a long time, but then a reed is ranked from the river in Nippur, and language undergoes a further metamorphosis. It is endowed with a power beyond that of our DNA, the ability to replicate across space and time, untethered from its host. It reveals itself now to be much more like a virus, gestating in the brain before spreading itself rapidly and widely wherever the symbols go, so that an idea a hundred or a thousand years old can enter and divert us, changing forever our trajectory, as I was once diverted, more than sixty years after his interment, by a single line of Yeats:

> We can make our minds so like still water that beings gather about us that they may see, it may be, their own images, and so live for a moment with a clearer, perhaps even with a fiercer life, because of our quiet.

So along with our genes, this is another thing we inherit and bequeath—our language. And what have your body and all its emissions been then? Not a mere husk, and neither an impediment. More like the soil in which grows the language of those we love. The frankest and freest product of the human mind is a love letter, says Twain, writing from the grave. Language sheds the body. Then only the words remain, in a book like this one, waiting for a rainy day, a comfortable chair, waiting for you, so they may be born again.

Acknowledgments

First off, a huge thank-you to Claire Gutierrez, Christian Lorentzen, and all the good folks at *Harper's Magazine* for believing people might want to know what the Romans did with their urine, to Kisha Schlegel for believing a person could write a whole book about bodily fluids, and to Gordon Shaw for telling me, late one night, over grappa, that he believed in me. Thank you to Ted Thompson and Jen Percy, whose advice helped guide this half-baked concept into a full-fledged book proposal; to PJ Mark, one of the sweetest and savviest folks I know; and everyone else at Janklow & Nesbit for bringing that proposal into the world; and to Peter Hubbard, Jessica Vestuto, Shelly Perron, Stephanie Vallejo, Sharyn Rosenblum, Maureen Cole, Tavia Kowalchuck, Erin Merlo, Jillian Perez, and the whole Mariner/HarperCollins crew for turning that rough idea into an actual book and for getting that book into your hands. Thank you to GWU's Department of English and the Jenny McKean Moore Fellowship for time and encouragement, and a big thank-you,

ACKNOWLEDGMENTS

with awe and respect, to Sally Butterfield for making sure every comma had a reason to be.

Thank you, as well, to the people whose friendship, guidance, and conversation were invaluable in the creation of this book: Whitney Peeling, Rawaan Alkhatib, Colby Somerville, Geoff Hilsabeck, Dave Eck, Eric Dougherty, Amity Quinn, Dylan Nice, Will Boast, Cara Blue Adams, Cam Terwilliger, John D'Agata, Jo Ann Beard, Catherine Imbriglio, Nancy Jacobs, Cristina Maria Cervone, Sam Abrams, Joe Fassler, Rachel Fagnant, Jonathan Vincent, Rachel Vincent, Brian Christian, Jenny George, Clarence Harlan Orsi, BG Cross, the dads of Baltimore, and all the folks who pulled me aside, after readings or in the middle of dinner parties, to tell me something about the human body. And a special, heartfelt thanks to Jesse and Stacey Butterfield, for getting my limp, lifeless body out of Soul Quest Ayahuasca Church of Mother Earth and into a queen-size bed at the Orlando Hilton.

This book wouldn't have been possible without the hard work of countless writers, scientists, scholars, and artists. To all the people who have been awed by the patterns of glycans on a mucin or the crystallization of a tear or a boyhood flatulence game in central Pennsylvania, thank you.

Thank you to my family: Mom, Dad, Liz, Hannah, Dante, Marcia, Bob, Daisy, Anaïs, Linden, Benj, Nick, Lynn, Dennis. And my other family: Margie, Nancy, Tony, Siobhan, Cavan, Gordie III, Janice, Lila, Gordie IV. And even to Jude. I wouldn't be here or be me without you.

And thank you to Erin and Teddy and Lula. Who in their right mind could ask for more.

About MARINER BOOKS

MARINER BOOKS traces its beginnings to 1832 when William Ticknor cofounded the Old Corner Bookstore in Boston, from which he would run the legendary firm Ticknor and Fields, publisher of Ralph Waldo Emerson, Harriet Beecher Stowe, Nathaniel Hawthorne, and Henry David Thoreau. Following Ticknor's death, Henry Oscar Houghton acquired Ticknor and Fields and, in 1880, formed Houghton Mifflin, which later merged with venerable Harcourt Publishing to form Houghton Mifflin Harcourt. HarperCollins purchased HMH's trade publishing business in 2021 and reestablished their storied lists and editorial team under the name Mariner Books.

Uniting the legacies of Houghton Mifflin, Harcourt Brace, and Ticknor and Fields, Mariner Books continues one of the great traditions in American bookselling. Our imprints have introduced an incomparable roster of enduring classics, including Hawthorne's *The Scarlet Letter*, Thoreau's *Walden*, Willa Cather's *O Pioneers!*, Virginia Woolf's *To the Lighthouse*, W. E. B. Du Bois's *Black Reconstruction*, J. R. R. Tolkien's *The*

Lord of the Rings, Carson McCullers's *The Heart Is a Lonely Hunter,* Ann Petry's *The Narrows,* George Orwell's *Animal Farm* and *Nineteen Eighty-Four,* Rachel Carson's *Silent Spring,* Margaret Walker's *Jubilee,* Italo Calvino's *Invisible Cities,* Alice Walker's *The Color Purple,* Margaret Atwood's *The Handmaid's Tale,* Tim O'Brien's *The Things They Carried,* Philip Roth's *The Plot Against America,* Jhumpa Lahiri's *Interpreter of Maladies,* and many others. Today Mariner Books remains proudly committed to the craft of fine publishing established nearly two centuries ago at the Old Corner Bookstore.